Missing Links

Missing Links

THE AFRICAN AND AMERICAN WORLDS
OF R. L. GARNER, PRIMATE COLLECTOR

Jeremy Rich

THE UNIVERSITY OF GEORGIA PRESS
Athens & London

A previous version of chapter 6 appeared in *African Historical Review* 40, no. 2 (2008): 62–83, and is used here by permission from the editorial board of the *African Historical Review* and Routledge, Taylor, and Francis.

© 2012 by the University of Georgia Press
Athens, Georgia 30602
www.ugapress.org
Designed by Walton Harris
Set in 10.5/14 Adobe Caslon Pro

Printed digitally in the United States of America

Library of Congress Cataloging-in-Publication Data

Rich, Jeremy (Jeremy McMaster)
Missing links : the African and American worlds of
R. L. Garner, primate collector / Jeremy Rich.
    p.   cm. — (Race in the Atlantic world, 1700–1900)
Includes bibliographical references and index.
ISBN 978-0-8203-4059-3 (cloth : alk. paper) —
ISBN 978-0-8203-4060-9 (pbk. : alk. paper)
1. Garner, R. L. (Richard Lynch), 1848–1920. 2. Primatologists—
United States—Biography. 3. Apes—Gabon. 4. Apes—
Collection and preservation—Gabon. 5. Apes—Collection
and preservation—United States. 6. Gabon—History—1839–
1960. 7. Americans—Gabon—Attitudes. 8. Racism—History—
19th century. 9. Africans—Public opinion, American. 10. Human-
animal relationships—Gabon. I. Title. II. Series: Race in the
Atlantic world, 1700–1900.
QL31.F65R53 2012
599.88096721—dc23   2011018020

British Library Cataloging-in-Publication Data available

For my sister, Melissa Rich (1967–2008),
and my friend James Conlon (1972–2009)

CONTENTS

ACKNOWLEDGMENTS

THIS BOOK BEGAN AS A HAPPY ACCIDENT. The kindness and support I received from so many unexpected places made me feel grateful as I stumbled about in my search for the frayed traces of R. L. Garner's life. First, the Smithsonian Institution's staff at the National Anthropological Archives and the Smithsonian Archives, especially Lorain Wang and Daisy Njoku, were extremely kind to me during my occasional forays into Garner's materials. The Wildlife Conservation Society kindly passed along scanned copies of Garner's correspondence with the Bronx Zoo curator William Hornaday. The volunteer staff of the Washington County Historical Society in Abingdon, Virginia, shared with me their knowledge of Abingdon. The Tennessee State Archives provided a valuable resource by sending Confederate records of Richard Garner's service in the Civil War. Father Gerard Vieria graciously allowed me to comb the Archives of the Congregation of the Holy Spirit, Garner's erstwhile nemesis, in Chevilly Larue, France. The Centre d'Archives d'Outre-mer in Aix-en-Provence also gave me valuable information about Fernan Vaz at the turn of the twentieth century. The Middle Tennessee State University Faculty Research and Creative Activities Committee provided the funding for my research in France.

Many people helped me as I worked on this project. Mark Blackman and Emily Pope-Blackman lent me a couch as I traveled to Gabon. I could not have accomplished anything in southern Gabon without the crucial aid of Alan Davis Poba, who helped introduce me to Garner's beloved southern Gabonese coast. I sincerely thank all of my informants in Ndougou, Ombouè, and especially the staff at the Mission Sainte Anne de Fernan Vaz. Father Louis Moto yet again welcomed the least likely member of the congregation of Notre Dame des Victoires de Glass back to the musty confines of the upper chamber of the church in Libreville. I hope no dead birds are lying in wait for the next visitor there.

While I struggled to find a way to domesticate R. L. Garner's unruly life into a manuscript, I thankfully had friends and colleagues to advise me. The University of Georgia Press has been very supportive of this project, and I particularly thank Derek Krissoff for taking a chance on it. Peer review can be a frustrating process, but the two anonymous peer reviewers understood the art of constructive criticism. Gregory Radick shared many ideas with me, and I hope he properly appreciates the absurdity of the existence of not one, but two Garnerians. My old friend and fellow *gabonard* John Cinnamon had to endure my long conversations about new Garner discoveries, as did my colleagues in the History Department at Middle Tennessee State University. Special thanks go to Bethany Hall for preparing the map for this book. Kairn Klieman gave me important insights on several draft chapters. Commentators at the Southeast Regional Seminar on African Studies and the Mid-America Alliance of African Studies helped sustain and sharpen my project, especially Margie Buckner and Jamaine Abidogun from MAAAS and Ken Wilburn, Owen Kalinga, Brett Shadle, and Nyaga Mwaniki at SERSAS. Regional Africanist interdisciplinary organizations are a joy and a vital part of Africanist research, even though they often do not receive the respect they deserve. Audiences at the annual African Studies conference at the University of Texas at Austin, the Southeast Africanist Network annual meeting, the Ohio Valley Historical Conference, the Tennessee Conference of Historians, and the Appalachian Studies Association challenged me as well.

My family paid a high price as I battered away on my keyboard. I hope the results will be interesting enough for them to actually finish reading the entire book. I am so lucky to have such a patient and insightful wife. Chantal, I thank you for feigning interest for so long in my supposedly glorious revelations about a long-departed primate collector. You must love me if you can put up with that. My mother, Kay; my father, Barry; my stepmother, Carol; and my mother-in-law, Carole, all offered their support as well. It is with deep sadness that I know my late sister, Melissa, will never turn a single page of this book, but our shared fondness of the unusual continues to lead me to surprising places. My beloved laughing and crying children, Lucien and Beatrice, came after the manuscript was largely completed. As children born in central Africa, they now inherit the tragic history of race that Garner actively participated in. May they learn to successfully overcome the prejudices that still resonate with Garner's caustic views of Africans all too well today.

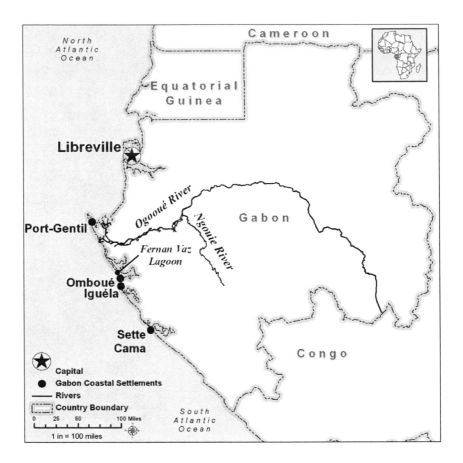

Cameroon

Equatorial
Guinea

Libreville

Ogooué River

Ngounié River

Gabon

Port-Gentil

Fernan Vaz
Lagoon

Omboué
Iguéla

Sette
Cama

Congo

South
Atlantic
Ocean

★ Capital
● Gabon Coastal Settlements
— Rivers
☐ Country Boundary

0   25   50   100 Miles

1 in = 100 miles

Missing Links

INTRODUCTION

ON 18 FEBRUARY 1911, U.S. PRESIDENT WILLIAM HOWARD TAFT amused himself at a special dinner held at the Willard Hotel in Washington, D.C.[1] The Gridiron Club, a social organization of prominent journalists, held its annual roast of the luminaries of the American economic and political elite. A score of senators and representatives rubbed shoulders with generals, Supreme Court justices, major newspaper publishers, railroad tycoons, the ambassadors of Austro-Hungary and Germany, and financier extraordinaire Andrew Carnegie. Attendees devoured a mountainous meal. Guests selected their choices from a crowded menu, which included steak, terrapin Maryland, crab *flakes en cassolette*, and a seemingly inexhaustible supply of alcohol — cognac, martinis, and a mysterious concoction known only as Gridiron punch. A server in a Cupid outfit, which apparently consisted of wings and "little else," handed each guest a witty valentine card.

After dinner, guests smoked cigars as they savored the night's manifold entertainments. A band in wigs played a series of songs in German, accompanied by a display of pictures of American life with faux-German captions, finishing with Theodore Roosevelt, "Schnickle Fritz." Members dragged out Speaker of the House Champ Clark of Missouri in chains. Comedians impersonated many of the guests and made speeches on their behalf. An ersatz Andrew Carnegie reminded the assembly that he wished to "die poor, but advertised." Victor Berger, the first Socialist Party member elected to Congress, received a bomb with a lit fuse as a gift. Racial caricatures made their way onto the stage as well. Supposedly, a Japanese spy had uncovered secret information on many of the congressmen in the room, which included "measurements of the honorable hole into which Congress throws a billion dollars every year." Because the club ensured that none of the guests' replies to the jokes would ever find their way into newsprint, the responses of the august attendees can be only a matter of conjecture.

An aging and portly Virginian named Richard Lynch Garner ascended to the spotlight amid this exclusively white, male, and exceedingly tipsy crowd. His origins in a small railroad town in southwestern Virginia and his poor finances hardly corresponded with the genteel company. Yet Garner had something that allowed him to enter such an august gathering despite having only a high school education. He had a chimpanzee. Several months earlier, Garner had returned in triumph to New York City with several apes from his travels in the rainforests of southern Gabon. Since 1892 Garner had traveled frequently to the French colony of Gabon to prove his theory that primate languages could be deciphered and made intelligible. Such assertions had made Garner a laughingstock of the American scientific community and a ready punch line for journalists in need of a monkey joke. However, the arrival of the chimpanzee Susie in America did receive attention in the press, thanks to Garner's assertion that he could converse with his African ward. Before the duo's show at the Gridiron Club, they had made public performances at the Bronx Zoo, the Philadelphia Zoo, and a number of lecture halls. Just as at their previous engagements, Susie stayed mute. Reporters noted that the audience enjoyed the animal's antics despite Garner's failure to elicit a few words from his fellow performer.

Primates had captured the imagination of many Americans in the early twentieth century, and Richard Garner had found a niche for himself that allowed him to draw crowds to his lectures. The Gridiron Club exhibition was just one stop in Garner's tours with Susie in 1910 and 1911, which introduced many Americans to his eccentric theories. Popular interest in Charles Darwin's theory of natural selection and the uses of evolutionary rhetoric by diverse groups in the United States during the Progressive Era opened the door for men such as Garner, who bucked the contemporary trend of armchair scholarship by conducting fieldwork in Africa. A rising demand in North America by zoos, museums, and many Americans to see primates allowed Garner to become one of the most well-known self-styled experts on Africa in the United States from 1910 until his death in 1920. Although primatologists have mocked Garner's methods and theories ever since he first emerged as a public figure in the 1890s, by 1910 Garner's ability to promote himself through his animals finally brought the Appalachian autodidact the acclaim he always felt he deserved.

The evolutionary theories so in vogue in Europe and North America at the turn of the twentieth century made Garner's career. Creatures that stood, or rather lumbered, in the imagined space between apes and humans teemed in the pages of novels and newspapers. Garner himself disdained these hulking hordes of fictional ape-men as mere fantasy, even as his efforts to animalize Africans and endow apes with feelings and languages blurred sharp divisions between humanity and the rest of the animal world. Garner's eccentric life embodied a different set of missing links. Wealthy directors of U.S. museums and zoos, colonial administrators, embittered African traders, and zoo visitors were all connected through Garner. His journeys weaved along Atlantic circuits of trade and knowledge, defying the traditional borders that still too often divide historians of the United States, Europe, and Africa.

Garner's life is a prime example of how discussions and research on human evolution were truly Atlantic in nature. Through the life and experiences of Richard Garner, this book explores the connections and opportunities created by Western demand for African animals in the early twentieth century. Neither his personal story nor the animal commerce that made him famous can be limited to one country. Understanding Garner's struggle to make himself a renowned expert requires an investigation into his conceptions of race, manhood, and Appalachian identity. In turn, Garner could only have found a platform for his unusual ambitions through three interrelated global events: the flourishing fortunes of North American zoos, the transatlantic demand for primates among audiences saturated by popular interest in Social Darwinism and evolution, and the subjugation of African environments and people in the age of high imperialism. French missionaries, officials, and anxious museum and zoo directors all entered into Garner's eccentric plans. Although scholars have long made brief references to the interrelations between colonialism, Social Darwinism, and Western consumption of exotic animals, they have never been explored together through the career of a single individual.

Garner's work was situated within a more extensive circulation of both knowledge and animals that had emerged in the sixteenth century with the expansion of European oceanic empires. This study builds on the rich literature of the past decade, which has exposed the transatlantic nature of scientific inquiry and intellectual life and which has explored concepts of weights, measures, and botany.[2] Although studies of Atlantic science are generally concerned with early modern history rather than developments after the mid-nineteenth century, researchers have rapidly recognized how evolution was deeply tied to circum-Atlantic intellectual, commercial, and political networks. Evolution provided a rich framework that could be appropriated throughout the world in a vast range of contexts—among, for example, Latin American intellectuals or New York bohemians or even New Zealand settlers.[3] Adrian Desmond and James Moore have forcefully argued that Charles Darwin's evolutionary ideas cannot be separated from his opposition to slavery and transatlantic debates over the nature of human races.[4] Garner's story brings together a different set of Atlantic threads severed by the end of the slave trade, as southern Gabonese communities who had relied on slave exports made alliances with a former Confederate soldier whose family had owned slaves. The circulation of specimens and knowledge from African to North American locales in the late nineteenth and early twentieth century has rarely been placed in an Atlantic context. Garner's case shows how older Atlantic political and trade networks survived the end of the slave trade, as well as how the new political and intellectual realities of African colonialism and the North American scientific community altered long-standing relationships.

Zoos in the United States, Garner's main clients, served to propagate evolutionary concepts and relied on Atlantic networks for animals. Garner's story links Atlantic connections and the existing historical literature on North American and European zoos and museums. Although historians have repeatedly noted how zoos promoted sharp distinctions along racial lines and celebrated a paternal vision of white professional management of African animals, few have examined the context of Atlantic commerce and colonial expansion that brought African primates to North America.[5] Gorillas brought zoos high numbers of visitors and put their institutions' names in the newspapers. William Hornaday, head of the Bronx Zoo in the early twentieth century, expressed a common opinion in 1915: "Undoubtedly the highest desire of every zoological garden and park, and of every showman, is to own and exhibit a real live gorilla of a size sufficiently large to compel both admiration and awe."[6] Live animals were prized, but very few survived the long trip by sea to transatlantic destinations. Garner's services thus had value for American zoos and museums, even if the men who managed these institutions found Garner's theories thoroughly flawed. In addition, the strengthening hold of French firms and the colonial regime allowed Garner access to animals at a time when Gabonese communities were losing control over the terms of exchange for their natural resources. Such conclusions correspond with dominant trends within the historiography of scientific research in colonial Africa, especially in southern Africa, in which state impositions of power and scientific research often worked hand in hand.[7]

No matter how much Garner endorsed biological racist views and white male dominance over Africans in his copious writings, he remained an outsider in the very scientific circles he wished so fervently to enter. Garner's public career had many obstacles, including his mountain South origins. The hills of eastern Tennessee and southern Virginia are often associated with evolution and scientific progress in the early twentieth century, but hardly in the way Garner would have liked. Coming from southern Appalachia hardly helped his cause, as the region was increasingly defined as an exotic and primitive backwater, out of touch with mainstream America.[8] U.S. and international journalists flooded into the southeastern Tennessee town of Dayton in 1925, five years after Garner's death, to watch high school biology teacher John Thomas Scopes try to defend himself against the charge he had illegally taught evolution in his classroom. Legendary iconoclast H. L. Mencken joined the crowd of observers and delighted in mocking rural Appalachians as a band of superstitious, near-simian buffoons. Eastern Tennessee represented a monolithic bloc of religious fanaticism dead set against intellectual freedom, save for a handful of enlightened unfortunates like Scopes himself. Scholars have dismissed such simplistic observations by contemporaries of the trial as a battle between scientific progress and simpleminded faith, particularly in the rhetorical duels between famed lawyer Clarence Darrow and former secretary of state and presidential candidate William Jennings Bryan.[9] Scholarship on Appalachia and southern religious history has noted how the Scopes trial marked a major victory for the rising fundamentalist movement in evangelical Protestant churches, both in the Virginia-Tennessee borderlands and throughout the nation.[10] Southern Appalachia remains in the historical literature as a region whose educational levels paled in comparison to the rest of the country and a place rarely associated with scientific innovation.

Garner went against the grain of many verities associated with the Scopes trial, even as his life was so closely tied to evolution and primates. Rather than view similarities between apes and man as an insult to God, Garner believed that Christianity was a set of lies that denied the truth of commonalities between man and other primates. Many prominent proponents of evolution in the 1920s, such as the famed American Museum of Natural History director Henry Fairfield Osborn, tried to reassure enlightened Christians that the drama of Scopes did not mean science and faith were always at odds. Osborn recoiled at William Jennings Bryan's declaration that the museum director fell in the same class of legendary atheists as Robert Ingersoll and Thomas Paine.[11] This accusation would have been a great badge of honor for Garner, who hoped to inter Christianity once and for all by toppling humankind from its supposed unique position. Evangelical Christians in the U.S. South argued that a firm paternal hand was needed to defend children's faith from dangerous ideas; Garner, by contrast, maintained that true southern white manhood could flourish only when the myths of Christianity were vanquished. Freethinkers enamored with evolution did exist in Appalachia in the late nineteenth and early twentieth century, in spite of easy generalizations about the piety of the people of this region. Some southern scientists and theologians accepted evolutionary theories in the late nineteenth century, but Garner's heady mix of free thought and primates did not fit with this more genteel tradition.[12] His story thus points out how little is known about dissenters from religious orthodoxy in southern towns who might have grasped hold of Darwinist ideas in ways less flamboyantly than Garner himself. A significant historiographic problem that Garner's life exposes is how the Scopes trial has obscured unconventional intellectuals who sought to appropriate scientific research and Darwinism in their own critiques of southern life.

Given how much Garner's career resonates with major themes in Atlantic and American cultural history, why has he remained such an obscure figure? Garner's absence from historical scholarship can be explained in several ways. First, his reputation as a self-deluded and thoroughly unscientific thinker hardly helped matters. Because Garner did not have any steady financial backing from any public or private institution from 1892 to 1911, he submitted a constant supply of essays on daily life in Gabon to U.S. magazines and newspapers to supplement his meager income. Unfortunately for historians of colonial Africa, few actually reached print. In addition, his papers have been almost entirely ignored since they were donated by his son Harry Garner to the Smithsonian. Garner's cantankerous tone might have put off publishers, but it could not have helped that hardly any Americans had ever heard of Gabon, besides a very small number of American Congregationalist and Presbyterian missionaries who had worked on the coast of Gabon since the 1840s.[13] The works of Robert Hamill Nassau, a Presbyterian pastor and doctor who lived in Gabon for more than three decades, drew some attention from African specialists such as the English traveler Mary Kingsley.[14] In a period where so many American denominations set up African missionaries, though, one small and declining mission community hardly distinguished itself from more prominent missionary efforts in Angola, Kenya, Liberia, Nigeria, and South Africa. Garner's reputation and his subject matter were surefire paths to obscurity.

To date, Gregory Radick's *The Simian Tongue* is the only work to seriously consider Garner's theories and place him in the context of changing scientific ideas regarding animal communication. As a historian of colonial Africa interested in how North Americans conceptualized African locales, I have meandered far from Radick's valuable contributions. Even where our paths do occasionally cross, we are headed in different directions. I am more concerned with how Garner's work articulated a white southern vision of the world and how his material became so overdetermined with meaning that it could be harnessed to serve wildly different agendas, such as Christian socialism and bohemian feminism.

Radick regards Garner as an unacknowledged pioneer of animal communication research, but the self-taught Virginian zoologist was a herald of other issues much more pertinent to African history and a global cultural history of animals. Although Garner at times spun hunting narratives that celebrated white manliness at the expense of African animals and people, like so many other writers in the early twentieth century, he also asserted that he was a caring father to monkeys and primates. Instead of having gorillas and monkeys symbolize African savagery, Garner praised his pets and contrasted their faithfulness and kindness to the cruelty of African people toward animals. These stories made him part of a dramatic shift in human-animal relationships in the twentieth century, in which Westerners made animals into objects of love and affection. Garner explicitly contended that empathy toward animals was a sign of white superiority. Garner's pioneering work in anthropomorphizing gorillas and chimpanzees helped pave the way for primate displays in American zoos, and directors such as William Hornaday consciously shared Garner's racist agendas. While later conservation efforts in colonial and postcolonial Africa did not explicitly discuss racial differences nearly as much as Garner did, the belief that Western environmental groups are best qualified to determine how African animals should be protected, with little consideration for African perspectives, is still very real.

Garner's African career also has value for historians of the Gilded Age and Progressive Era in several ways. His entire life stands in contrast to common stereotypes about southern Appalachia in the late nineteenth and early twentieth century. Missionaries, travel writers, and some Appalachian townspeople claimed in the late nineteenth century to have discovered that the mountain South was far removed from the social and intellectual development of the rest of the country. Garner's later life was dedicated to removing himself from this baggage. Instead, he endorsed the same ideas of southern uplift through scientific development that so shaped politics and intellectual life in the New South. Garner viewed scientific achievements as the means of transcending his middle-class background and the heritage of southern defeat. Garner also felt he had much to teach northerners about race and science, even if few others turned to chimpanzees as props to prove their points. Although Garner's atheism hardly made him typical for southern intellectuals in the early twentieth century, he had a thoroughly conventional perspective of the intellectual inferiority of people of African descent. Garner agreed with practically every aspect of what historian Idus Newby defines as "race orthodoxy" among white middle-class southerners in the early twentieth century: unchangeable biological racial differences, the intellectual inferiority of black people, the need to ensure that political power remained only in white male hands, and the belief that white southerners understood people of African descent better than northerners.[15] Garner drew from his knowledge of animals and daily life in southern Gabon to support his allegations. While many white southerners of various political persuasions conjured up images of savage Africa to defend lynching and Jim Crow laws, Garner actually had lived in an African colony.

Although Gabon was among the least familiar regions in Africa for American audiences—a situation that has not drastically changed a century later—Garner's depictions of African animals and people did resonate with debates about race, class, and empire in the Progressive Era. From Garner's first visit to Gabon in 1892 to his final sojourn there during World War I, the United States had established an international empire in the Pacific Ocean and the Caribbean. Garner himself may have been as thoroughly convinced of white superiority as those who claimed Filipinos could not govern themselves, but he contended racial differences made colonization a doomed enterprise. Eric J. T. Love has contended that a long intellectual tradition among middle-class Americans brought together white supremacy and opposition to foreign colonial expansion, a mindset that advocates of U.S. global expansion managed to overcome only in the 1890s.[16] Garner belonged to this racist opposition to American imperial expansion. Garner also hoped to provide evidence for another long-standing use of African images in U.S. political discussion: the claims that knowledge of Africans supported the resettlement of African Americans in Africa. He was not alone; southern eugenicists like Earnest Sevier Cox visited West Africa to provide evidence to support African American resettlement schemes, later endorsed by politicians such as Mississippi senator James Theodore Bilbo.[17]

Garner's argument is an example of what Amy Kaplan has described as the "anarchy of empire," in which various Americans could pursue radically different ends in discussions of U.S. empire at the turn of the twentieth century.[18] An African career Garner began at middle age opened the door to new possibilities far removed from high schools in eastern Kentucky or real estate speculators' offices in central Virginia. As Harilaos Stecopoulos has recently noted, distinctively southern engagements with empire have been largely ignored in the growing literature on U.S. empire studies.[19] Neither has the new cultural turn of American engagements with the world in the early twentieth century seriously considered Africa, despite the notable exceptions of Jeannette Eileen Jones and Ibrahim Sundiata.[20] Descriptions of African locales by American travel writers could serve a host of different ends, even if they conveyed similar messages of racial differences.[21] Freethinkers such as Mark Twain could join pious Presbyterians in imagining the passive sufferings of Congolese people. While Theodore Roosevelt led a safari in British East Africa to celebrate a virile form of

Anglo-Saxon masculinity, Roosevelt's elderly associate Charles "Buffalo" Jones sought to lasso big game rather than kill them out of a sense of guilt over the destruction of animal life in the American West. African American clergymen such as Henry McNeal Turner and Pan-Africanist activist Marcus Garvey both promoted the emigration of people of African descent to Liberia, even if their views on Christianity and politics did not always agree.[22] Garner does not appear in Jeannette Eileen Jones's exploration of varied American representations of Africa in the late nineteenth and early twentieth century, but his story fits well with her contention that naturalists such as Carl Akeley imagined African environments and animals menaced by modernity and African people.[23]

Besides contributing a new understanding of how Africa played into American discussions of empire, this work seeks to provide a more contextualized perspective to the rich veins of scholarship on European and North American travel writing on Africa. Simply put, there is far too little effort made to use travel texts as a means of recovering African agency. Perhaps the most glaring example of this lacuna can be found in the voluminous writings about Garner's contemporary Mary Kingsley, who visited in Gabon in 1895.[24] In all the many works that analyze the celebrated English author's career and performances of gender, nary a single author actually bothers to ground Kingsley's narratives in the economic and political shifts that French colonial rule brought to Gabon. By contrast, this study places Garner's writings in a Gabonese context. Garner observed how enthusiastically he was greeted by southern Gabonese communities, and he interpreted their support as evidence that they regarded him as their intellectual superior. One could simply look at what Garner's writings exposed about Western conceptions of Africa, as did Mary Louise Pratt in her seminal deconstruction of Franco-American traveler Paul Du Chaillu's accounts of the same Gabonese rainforests that Garner later visited.[25] Yet the very earnestness of such projects in undercutting Western ethnocentrism can ignore what such texts might expose about the motivations and worldviews of Africans. It is easy enough to deny the subaltern can speak when one simply treats any particular African context as just a blank slate.

Although Garner constantly denigrated Africans, a close reading of writings reveals how dependent he was on African hunters, traders, and workers. Southern Gabonese communities enlisted Garner into their own struggles with the encroaching French colonial state and the brutal exploitation of concessionary companies. Garner's unpublished records denote the negative results of French colonial rule in the concessionary period. Colonial guards stripped impoverished villages of merchandise and animals. Administrators arbitrarily razed towns, jailed Gabonese people, and cut deals with concessionary companies to defraud Africans. Because he lacked formal support from the French government and could not call on the spiritual and financial capital available to missionaries and concessionary companies, Garner was an isolated figure. Unlike scholars in Europe and North America, Garner actually conducted research in the field for more than two decades. Furthermore, he desperately needed African expertise to provide him with knowledge of primates as well as of live animals, and so Gabonese people had the necessary leverage to pressure Garner to aid them. Because the American had only a distant relationship with the colonial government from the early 1890s until World War I, Nkomi people alternately flattered and threatened Garner to defend their interests.

Garner's expeditions to Gabon were thus more complicated than a mere story of exploitation. Southern Gabonese communities had long sold slaves, ivory, and rubber to visiting Portuguese and other European traders. Nkomi, Gisir, and Orungu people in coastal Gabon placed Garner in a long-standing tradition of African hosts and European guests who had coalesced over the course of Atlantic slavery. From the 1880s onward, these traditions of Gabonese control over foreign visitors had become severely undermined by the French administration. Garner transplanted his idealized view of paternal white supremacy in the United States to Africa. He and southern Gabonese people developed an uneasy and shifting set of alliances that sometimes worked together against the incursions of the French colonial government. Eventually, Garner relied extensively on Gabonese knowledge of animals, even as he upheld his belief in white intellectual superiority. His admiration for African knowledge denotes how he joined a range of other foreign scientific researchers in the nineteenth and early twentieth centuries who drew from African collaborators, even if these indigenous partners rarely received their due in published work.[26] Much as in the case of the eighteenth-century French botanist Joseph de Jussieu in Spanish America, Garner's unusual position as an independent and vaguely anticolonial operator allows for an understanding of actors who might be obscured by more conventional accounts that favor state and institutional connections between colony and metropolis.[27] Garner's negotiations also provide new insights on how central Africans struggled to guard their political and economic autonomy during the era of rapacious exploitation by concessionary companies. They underscore how coastal African Atlantic communities struggled to maintain older understandings that had developed in the slave trade of Europeans as guests who respected the authority of indigenous leaders.

An investigation of Garner's negotiations with Gabonese people gives new perspectives on animal collecting in the late nineteenth and early twentieth centuries. Most studies of the rise of zoos certainly highlight the role of colonial governments in facilitating the sale of animals. However, these studies have concentrated more on the European and North American aspects of the exotic animal trade. This business may never have become a major part of colonial African economies, but dealers did have to negotiate with both colonial governments and indigenous communities. Their activities can furnish insights on how various colonial governments tried to regulate hunting and land use, as well as how African traders adjusted to European administrations. Africans can be understood as historical actors in their own right, instead of as passive victims or shadowy suppliers to be ignored. My book thus corresponds to recent work on Atlantic histories of science that seek to recover, however imperfectly, the crucial role of indigenous Native American and African informants in the production of Western scientific knowledge.

Even as he dreamed of becoming a famous scientist, Richard Lynch Garner hardly had an impressive pedigree, especially by the standards of the Gridiron Club. He was the sixth of ten children reared by Samuel and Margaret Garner.[28] They constituted a solidly middle-class family in the 1840s and 1850s in Abingdon, a town located close to the Tennessee border in southwestern Virginia. Samuel Garner ran a foundry and a general store. In the early 1850s the construction of the Virginia and Tennessee railroad that connected Knoxville to Richmond and the East Coast made Abingdon a prosperous town. Garner's family reaped the benefits of improved economic connections between their home and the rest of the country. The Garner family owned several slaves, and Samuel's business furnished supplies to farming and mining enterprises that relied on slavery. Like many other well-off families in southern Appalachia, Samuel

Garner supported secession rather than share the Unionist sentiments of poorer farmers from more isolated regions in southwest Virginia and eastern Tennessee. On the dividing line between town gentry and poor rural households that Appalachian historian David Hsiung has noted, the Garners identified with economic growth and the defense of slavery.[29] Shortly before the war began, Garner's family moved to the nearby town of Bristol, Tennessee, where some of his sisters still lived four decades later.[30]

This relatively comfortable life came to an end with the Civil War. Most townspeople in Abingdon rallied to the Confederacy. Although rural farmers in the east Tennessee and Virginian mountains did not generally back the rebellion, townspeople who shared the sentiments of the political elites of southern states rallied to the cause. Samuel Garner served as an officer in a cavalry regiment. Richard Garner enlisted at fourteen years of age in 1862 and later was captured by Federal troops. He escaped a prisoner-of-war camp in Baltimore and then rejoined the Third Tennessee Mounted Infantry Regiment on 22 June 1864. Despite his age, Garner fought against Federal troops in southwest Virginia at the battle of Bull's Gap and Morristown. In a 1916 letter, Garner recounted his military service and his capture at Paperville, Tennessee, on 13 December 1864, after a Michigan Federal regiment defeated Garner's unit under the command of Alfred J. Vaughan.[31] During Garner's service, southwest Virginia was torn apart by fighting between roving bands of Confederates, Federal troops, and deserters from both camps.[32] After Federal troops captured Abingdon immediately after Garner's own surrender and burned part of the town to the ground, Garner was held prisoner for several months.

Garner's negotiations with the dislocations of the Civil War and Reconstruction period are hard to piece together. He attended Jefferson Academy for Men in Blountville, Tennessee, from 1865 to 1867, but then chose to move west for a number of years. At least according to a magazine article, Garner battled Apaches in northern Arizona and rode broncos in Colorado and Texas.[33] Oddly, given his penchant for long-winded autobiographical essays, Garner remained very quiet about this period of his life. Garner did make several references to living with Native Americans, and one early story he wrote in 1871 suggests he held them in far greater regard than African Americans.[34] By the mid-1870s, Garner had returned to southern Appalachia and found work as a schoolteacher in various parts of eastern Tennessee and Kentucky.[35] In 1872 he married Mary Gross, from Lockport, Kentucky.[36] They had one child, Harry E. Garner. During the 1870s and 1880s, Garner traveled frequently and eventually relocated to Roanoke, Virginia. This city's economy flourished in the mid-1880s, and Garner tried to profit from the city's rising fortunes as a regional industrial center by becoming a real estate speculator.[37]

Garner sought to insert himself into national scientific discussions rather than remain mired in teaching and real estate. Popular interest in evolutionary theories and the rise of new technologies provided him with the means to follow his dreams. At a phonetics conference, participants criticized Garner for daring to suggest that animals might have their own primitive languages. He observed some monkeys in a Cincinnati zoo in 1884 and became convinced that one could unlock the mysteries of animal communication. Garner read Darwin, linguist Max Müller, and other noted scientists. And the opposition he encountered only strengthened his resolve, as befitted a freethinker who believed that dogmas always stood in the way of individual manly achievement: "I could not find no literature on the subject [of animal speech] except the negative ipsi dixits [sic] of those microscopic authorities who have filled whole libraries with sophistry. . . . The goal at which I was aimed was not obscured by the universal negation and derision that I had to combat."[38] Once Garner became aware of Thomas Edison's wax cylinder phonograph, he immediately recognized how the new invention could support his own research. By using the machine to understand animal communication, Garner also could claim to be a technological pioneer whose efforts deserved an international audience.

Garner also tried his hand at literary efforts that expressed his interests in technological change and Appalachian folklore. Although these productions drew practically no attention in comparison to his animal-communication research, they did give an early indication on how Garner sought to present himself as a specialist. *Nancy Bet and Other Stories*, a collection of stories he privately published in 1891, was largely written in southwestern Virginian dialect.[39] As Katie Algeo has noted, townspeople from southern Appalachia joined northern travel writers in producing "local color" stories that portrayed the region as an exotic, superstitious land far removed from the American mainstream.[40] Garner's own narrative voice was written in grammatically correct English, which he used to distance himself from the popular image of Appalachian primitive culture so in vogue in the late nineteenth century. Although Garner did try his hand at florid poetry lauding the Confederacy, he also wrote a novel that explored how women found office jobs in Virginia, thanks to the advent of the typewriter. Garner's goals of describing and explaining scientific innovations as well as the customs of people unfamiliar to his readers continued in his work on Gabon.

Beginning in the late 1880s Garner quickly learned how to use the popular press to promote his career, even as formally educated scientists snubbed his public experiments. Eventually, he managed to secure some support from luminaries such as Thomas Edison through individual requests. He found an ally who could promote a public persona of manly adventure that could also attract support: literary impresario Samuel McClure, editor of *McClure's* magazine. McClure purchased hundreds of essays and stories from a dizzying list of clients that included Robert Louis Stevenson, Rider Haggard, and Rudyard Kipling. After buying the writers' work, he then sold it to hundreds of newspapers. While McClure's syndicate made him a literary legend, between 1884 and 1892 he struggled to find the funds to continue his ambitious operations.[41] Why McClure agreed to give Garner a chance is not entirely clear, but McClure's determination to sell sensational stories of adventure certainly helped Garner's cause. McClure launched an extensive public relations campaign that included numerous letters to prominent authors and scientists. Garner's novel theories and his sudden willingness to travel to far-off central Africa to prove them made the Virginian an overnight sensation, particularly since McClure aggressively marketed his client to dozens of newspapers.

The year 1892 was a watershed in Garner's life. Through the backing of various patrons, he published his theories in leading magazines and authored his first book, *The Speech of Monkeys*. Its provocative claims that animals spoke languages that could be deciphered through observation and phonographic recordings sparked controversy and sensation in England and North America. In the same year, Garner announced he would bring an elaborate set of electrical equipment and a steel cage to southern Gabon, the same region where Franco-American naturalist Paul Du Chaillu had claimed to have killed a gorilla in 1859. From the fall of 1892 Garner stayed in Gabon for more than a year.

The next two decades would be far less kind to Garner's public image than the tremendous support he had first received. When he admitted that his findings did not allow him to become proficient in the language of primates, the press largely turned on him. He secured funds from some Chicago investors to return to Gabon in 1894 and again stayed there for more than a year. While abroad, the English newspaper *Truth* further damaged his career by claiming that Garner had falsified his research and had not stayed alone in the jungle. Garner returned to England and then the United States to discover that he had become a laughingstock. Distraught, he headed back to Gabon in 1900 and stayed until late May 1902. Garner tried with little success to restore his reputation for the next two years. He finally returned to Fernan Vaz in 1904, and he remained there for the next five years, despite his depleting resources. Nearly destitute, Garner left Gabon in 1909 with his pet chimpanzee Susie and an undying hope in his redemption as a scientist.

His fortunes recovered significantly from 1910 until his death on 25 January 1920. Garner held numerous lectures with Susie and claimed he had learned dozens of words in the chimpanzee's language. While scientists remained skeptical, Garner's adventurous spirit and his knowledge of central African animals caught the attention of William Hornaday, curator of the Bronx Zoo and a nationally respected advocate for conservation. Hornaday hired Garner to acquire apes for the Bronx Zoo on expeditions in 1911 and from 1912 to 1914. The Virginian managed to bring two different gorillas back alive to New York—a feat that had been accomplished only twice beforehand. Then Garner returned to America in the fateful summer of 1914. He stayed in New York for the next three years, struggling to find the necessary patronage to revisit his former haunts.

He succeeded. A Philadelphia business owner and big game aficionado, Alfred Collins, agreed in 1916 to join Garner on a new expedition sponsored by the Smithsonian Institution. World War I prevented Collins from accompanying Garner, but the Smithsonian sent young taxidermist Charles Aschemeier (1893–1973) to Gabon with the aging researcher. This unlikely duo hunted for specimens from the spring of 1917 through March 1919 and sent more than two thousand animal specimens to the Smithsonian. During the war the museum lacked the funds to pay for their return home, but when the men finally arrived in America, Garner found that he had redeemed himself. Major newspapers and magazines such as the *Century* and the *Forum* published Garner's articles. Recognizing that his age precluded another Gabonese expedition, Garner convinced a New York attorney to establish a center in Florida for the study of primate and monkey communication. Garner stopped by to see family members in Bristol, Tennessee, for the Christmas holiday in 1919 before heading south. En route for the new project, complications from a kidney disorder finally took Garner's life while he stayed at a hotel in Chattanooga. One of the most colorful southern public intellectuals of the Gilded Age and Progressive Era had passed away.

Sources on Garner's life pose challenges for interpretation. Shortly after Richard Garner's death in 1920, his son Harry donated his papers to the National Anthropological Archives of the Smithsonian. They constitute the main set of sources about his life in Gabon, along with his numerous publications and newspaper articles. These materials consist of more than one hundred unpublished essays, correspondence with his son, and a diary kept from 1905 to 1906. Unfortunately, the collection contains very little written information on Gabon before 1904, and Garner's reticence regarding his life prior to his emergence as a public figure in the early 1890s makes his early life difficult to trace at times. Garner's correspondence with Hornaday from 1911 to 1917 and the Smithsonian for his 1916–1919 expedition are also very important for understanding the last ten years of his career. Like many other Western travel writers, Garner often was frustratingly vague in regard to geographical locations and dates of events discussed in his writings.

To contextualize these materials, I consulted colonial administrative archives from Fernan Vaz kept at the Centre d'Archives d'Outre-mer at Aix-en-Provence and Catholic missionary archives stored at the Congregation of the Holy Spirit's archives in the Parisian suburb of Chevilly Larue. These archives contain only a handful of references to Garner, but they are the most valuable repositories of writings for reconstructing daily life in southern Gabon from 1890 to the early 1920s. These letters and reports are supplemented with more than a dozen interviews conducted in December 2008 and January 2009 with elderly Nkomi and Gisir individuals living in the Fernan Vaz district. Although this limited sample uncovered no reference to the American researcher, informants did furnish numerous details on politics and cultural practices in early twentieth-century southern Gabonese communities. Through references in French archives and the Gabonese interviews, it is possible to partially recover Gabonese perspectives on foreigners and French colonialism that are not always apparent in Garner's own writings.

Rather than give a chronological review of Garner's life, I have organized this book thematically. There are several reasons behind this decision. Very few sources have survived on the first four decades of his life. Furthermore, his extremely broad range of interests on Gabonese fauna and people defies a typical biographical approach. A study set up along chronological lines would not adequately capture his engagements with Atlantic intellectual and economic networks or do justice to the nexus of cultural and social issues he engaged with. Thus, chapter 1 contains an overview of the southern Gabonese coast from Garner's arrival in 1892 to his final departure in 1920, a region as unfamiliar to most African specialists as it is to scholars of American history. Along the southern Gabonese coast, African communities that had once profited greatly from slave trading with Portuguese, Brazilian, and São Tomean merchants had to adjust to the end of transatlantic slavery and the gradual encroachment of the French naval government. French officials incorporated the southern Gabonese coast into the colony of Gabon in the 1880s and early 1890s but lacked the resources to completely reshape older Gabonese political institutions. In 1899 the French government established a system of concessionary companies modeled on the notorious Independent State of the Congo ruled by Leopold II of Belgium. Widespread misery, famines, and declining economic fortunes were the end results. Poll taxes and the advent of World War I augmented the hardships most southern Gabonese people faced. Chapter 2 explores how Garner negotiated with North American backers, the French government, and Gabonese people to obtain funds and animals in this troubled setting. Although the excesses of the concessionary system appalled Garner, he received privileges from the French government to act as an independent trader. Museums and zoos slowly became willing to purchase Garner's specimens and after 1910 finally chose to back major expeditions on his behalf.

The next three chapters discuss Garner's various positions in Atlantic scientific and cultural life, from Africa to North America. Chapter 3 reviews how Garner and his Western critics sparred over whether or not Garner met the requirements of heroic manhood, as well as Garner's own personal failures to maintain his ideal of a mutually supportive family. Garner managed to reconcile with French and English residents of Gabon

initially put off by his unusual behavior, but he could acquire support
from museum and zoo directors only after almost two decades. Instead of
dwelling on the Lost Cause, Garner instead argued that individualism was
the sole way that one could truly be a male hero. Naturally, Garner downplayed his reliance on Gabonese people for animals and labor in his self-
representations. Chapter 4 examines Garner's efforts to promote himself
as an expert on racial matters. Although Garner agreed with most of the
dogmas promoted by Madison Grant and other leading racist American
theorists, he also recognized the value of Gabonese knowledge of animals
and plants. His ideas on race thus amalgamated concepts borrowed from
the free-thought movement, racist objections to American colonial expansion, typical white southern bourgeois views on African Americans,
and popular understandings of Social Darwinism in North America.
He fashioned for himself a persona of a caring white patriarch oversee-
ing the true interests of people of color, which led him to write a series
of blistering critiques of French colonialism. Eventually, Garner chose to
endorse French rule, largely because of his vituperative dislike of most
Gabonese people and because the colonial state allowed him to operate
in Gabon.

Animals take center stage in Chapter 5. Garner rejected Christian doc-
trines regarding the special position of humanity, and so believed that animals could indeed have emotions and display affection. In keeping with
the growing conception of animals as becoming part of human families,
Garner wrote and displayed photographs in which monkeys and primates
acted as his surrogate children. Against this ideal of wise white human
parents and animal children, African people appeared as vicious and callous threats to animals and white people alike. Garner could never have
constructed animal families without Gabonese people killing the parents
of his menagerie, yet Garner still resolutely put the culpability for animal cruelty solely on his Gabonese trading associates. At the same time,
Garner also savored his dog's violent behavior towards his Nkomi neighbors and enjoyed the spectacle of primates owned by European residents
of the colony. Domesticated animals in Gabon did not serve as signs of
middle-class civility but rather as a means of degrading Gabonese claims
of equality with foreign residents of the colony.

The final chapters reflect on how Gabonese and American audiences
responded to Garner. Chapter 6 considers how Nkomi, Fang, Gisir, and

Orungu clans asserted their rights as the true masters of their domains in their dealings with Garner. Although Garner certainly enjoyed his incorporation into Gabonese communities through healing rituals and praise for his exploits, he tried to avoid reciprocal agreements to pay and protect Gabonese people in return for their support. While the American researcher presented himself as always in a position of command, Gabonese people employed Garner as a source of income and as a mediator with French officials. Garner's narratives also expose the rapidly changing understandings of spirituality in southern Gabon in the early twentieth century, as his Western technology became another set of power objects in an already crowded field of mission and indigenous spirituality.

In chapter 7 the discussion moves from Gabonese views on Garner to the varied reconfiguring of Garner's work in the English-speaking popular press. Garner's declarations regarding primate communication were viewed as so fantastic that they naturally lent themselves to satire and social commentary. Observers as far as New Zealand suggested that Garner's research conclusively placed people of color out of the same species as white people. Still others developed complicated tales of race and politics, where Irish American Democrats, white Americans pretending to be Japanese students, and chimpanzee suffragettes could come together. Gabonese gorillas exhibited at the Bronx Zoo between 1911 and 1915 served to celebrate white professional expertise through providing care for bewildered and vulnerable creatures. Yet female artist Eugenie Shonnard and the young modernist writer Djuna Barnes suggested they themselves had a female bond that transcended the boundaries of species and the narratives imposed by the zookeepers. Just as Gabonese people tried to force Garner to follow their own spiritual and political conventions, Western audiences edited out Garner's own individual agendas to serve their own interests. In one of the many incongruities of Garner's work, he would be removed from many of his own narratives in the same way Garner had downplayed the agency of Gabonese people.

Unfortunately, the following century has seen little change in the ways southern Gabonese communities have lost control over how they and their lands have been portrayed and governed. Petroleum and timber exports controlled by foreign companies dominate the economy, and the postcolonial state has long made clear that southern Gabonese people will have little say in how natural resources such as animals will be managed. Garner's old stomping grounds became the site of the American television show *Survivor* in 2008. Contestants unknowingly lived out Garner's dream of bringing American tourists to the southern Gabonese coast so that they could test their mental and physical toughness. Local people received little benefit from the performances, just as Gabonese people figuratively and literally lost out in Garner's trade for animals. This book is thus a genealogy of unequal relationships between Africa and the United States, rather than just a narrative that ended at Garner's death.

CHAPTER ONE

The Southern Gabonese Coast in the Age of Garner

RICHARD GARNER SPENT MOST OF HIS TIME IN AFRICA living along the Fernan Vaz (Eliwa Nkomi) Lagoon on the southern coast of the French colony of Gabon. Over time, he traveled through the grasslands and rainforests along the Rembo Nkomi River, which empties into the lagoon from the east. To the south of Fernan Vaz, the flat plains and forests of the Mpivia River and the jungles surrounding the Ngovè and Sette Cama Lagoons became equally familiar to Garner through his travels in search of chimpanzees, gorillas, and other animals. Garner briefly lived along and regularly passed through the tangled web of small rivers that make up the Ogooué River Delta, the largest waterway in Gabon, located roughly a hundred miles away from Fernan Vaz. He also had to regularly visit the French trading and administrative center of Cape Lopez (now Port-Gentil). Before 1920 roughly five hundred Africans lived in this small town with several dozen Europeans, largely consisting of French administrators, British and German traders, and a scattering of hard-drinking Norwegian whalers. Despite how unimpressive Garner found the colonial capital of Libreville on the northern Gabonese coast, whose eight thousand or so African residents made it the largest town in all of Gabon, legal and medical matters occasionally forced him to go there.

This chapter investigates the natural and social environment of southern Gabon, a region that had been transformed by Atlantic slavery and other transnational commercial ties for centuries before Garner's arrival. Slave trading had constituted the main occupation of most prominent

Gabonese traders until the mid-nineteenth century, when the decline of slave plantations in the Americas and the Portuguese colony of the São Tomé and Príncipe islands led southern Gabonese people to collect rubber and ivory to sell to English and German traders. Southern Gabonese people had developed a form of diplomacy that welcomed foreign guests like Garner as long as they accepted the sovereignty of local people. French colonial officials slowly forced southern Gabonese people to abandon these older traditions in favor of surrendering their independence but lacked the means to conclusively destroy local political institutions before World War I.

Garner became linked to recent and older Atlantic networks in Fernan Vaz. Franco-American traveler Paul Du Chaillu made southern Gabon famous among European and North American scientists for his claims of seeing and shooting live gorillas there. The first gorilla skull had reached the United States via an American visitor to Gabon in the 1840s; he had purchased his prize from a Gabonese slave hunter.[1] Just as in so many other encounters between North America and Africa, Atlantic commerce had led to the movement of knowledge and scientific evidence between the two continents. Gabonese and Westerners alike recognized the link between trade and science.[2] Du Chaillu and others received aid for their gorilla hunts in the name of scientific progress from Gabonese leaders seeking to develop commercial partnerships. Garner and Gabonese people both consciously linked each other to these older relationships, even as French officials and companies asserted their control over Gabonese natural resources.

Geography and Society in Southern Gabon

Garner loved the Fernan Vaz Lagoon. "At every turn nature offers a lovely surprise. Trees glow with flowers of many colors; and one tree bears great clusters of pendulous violet-colored bloom in a profusion I have never witnessed elsewhere. Indeed, all along the banks of lakes and rivers scores of flowers grow in the greatest possible profusion of form and color," he rhapsodized shortly before his death.[3] The diversity and richness of this region's animal life allowed Garner to eke out a livelihood in central Africa.[4] After Du Chaillu had chosen to make Fernan Vaz his home in the late 1850s and mid-1860s, European and North American readers learned of the large numbers of wild buffalo, chimpanzees, and gorillas living close to the lake.[5] Romanian princes, German naturalists, English sea captains, French administrators, and missionaries organized hunts along the three lagoons of Fernan Vaz, Ngovè, and Sette Cama in the late nineteenth century.[6] Nearly a century after Garner last set his eyes on the plants and animals that so inspired him, the Gabonese government declared in 2002 that Ngovè and Sette Cama would henceforth be national parks.

Southern Gabonese environmental conditions did not always leave Garner happy, especially during the main rainy seasons between August and December and again between late February and May.[7] Downpours soaked Garner on long canoe rides, because it took several days just to go from Cape Lopez to the administrative post of Ombouè in Fernan Vaz. Driver ants, mosquitoes, termites, and the dreaded itchiness that followed bites from *fourou* flies annoyed Gabonese and foreign residents alike. More importantly, the high iron content of the region's soil limited agriculture in the region. The American had to rely on his African neighbors for food, despite his own fields of sweet potatoes and cassava. Garner's willingness to live isolated from other Westerners left him cut off from supplies. Last but not least, Garner's frenetic displacements on these waters to hunt and to acquire provisions required merchandise and money to pay canoe workers.

Foreigners had to negotiate with Africans if they wanted to make a living along the Fernan Vaz Lagoon and other coastal regions in the late nineteenth and early twentieth century. Those who lacked the financial and spiritual resources of Catholic missionaries or the command of French gunboats had to negotiate with their African hosts as relative equals. These deals were rife with misunderstandings. Indigenous social and political institutions often confused Garner, much as they did other Westerners who chose to reside in southern Gabon. Few visitors developed enough competence in Gabonese languages, particularly the coastal lingua franca Omyènè, to grasp the complex politics and spirituality of local people. Garner himself never claimed to master Omyènè, although he characteristically claimed African languages were far more primitive than European tongues.[8]

One major point of confusion came from contrasting European and Gabonese conceptions of ethnicity. Various travelers, as well as French administrators and missionaries, drew on prevalent constructions of race and linguistic affiliation to impose clearly separated categories of ethnicity on the diverse communities in southern Gabon.[9] Such artificial ethnicities did not correspond well with how Gabonese people developed bonds of solidarity with one another. Above all, the clan constituted the foundation of Gabonese politics. People speaking different languages claimed a common heritage by belonging to the same clan. Conversely, Fang-speaking migrants entering Fernan Vaz in large numbers from the north did battle with one another despite their common language, family structure, and spiritual beliefs. Villages usually belonged to a single clan, led by councils of men who held spiritual and political authority. The decentralization of power in the thinly populated region frustrated traders and administrators alike, because nearly every village acted as an independent state.

Coastal Omyènè-speaking Nkomi clans had a very long tradition of hosting foreign visitors, for reasons Garner knew intimately from his childhood in antebellum Virginia. Since the sixteenth century, canoes had loaded slaves from the sandy banks of the Fernan Vaz Lagoon and the Ogooué Delta to European vessels anchored offshore.[10] Compared to the millions of Africans shipped from farther south on the Congo River to the Americas, southern Gabon became an important supplier of slaves only in the early nineteenth century. When British naval vessels began to patrol the North Atlantic coast of West Africa in search of slave ships after the Napoleonic Wars, traders from Brazil, Spain, and Portugal found the relative obscurity and the tangled rivers a perfect locale to avoid detection. Clients from the Portuguese island colonies of São Tomé and Príncipe, only several hundred miles off the Gabonese coast, also bought slaves from Nkomi traders. A decade before Garner first came to Gabon, São Tomeans kept purchasing captives on the Ogooué Delta and Fernan Vaz, despite the official abolition of slavery in all Portuguese colonies in 1875.

Nkomi clan leaders considered Garner to be their client and favored guest, following the policies of the most powerful political leader in Fernan Vaz in the nineteenth century, Oyembo Onanga (reigned ca. 1840–83). Paul Du Chaillu, along with Portuguese slave traders and English and German merchants who moved into Fernan Vaz after the 1860s, all depended on Oyembo Onanga's protection. His power resided in his position as *rengondo*, the supreme leader of all the Nkomi families, and his esoteric knowledge of supernatural forces, particularly powerful *ombwiri* water spirits said to reside in the lagoon. Throughout the twentieth century, the memory of Oyembo Onanga remained synonymous with brutal executions, slave trading with Europeans, and his command over trade and people in his domains.[11] Contemporary descriptions by Du Chaillu and other visitors tended to be less impressive than stories told by Nkomi people, as the aged king struggled to maintain his control over Nkomi clans and trade in ivory, rubber, and slaves passing from the southern Gabonese interior to the coast through Fernan Vaz. The *rengondo* chose to develop close partnerships with European traders and even the French government, perhaps to offset the threat of rebellions. In 1858 he appointed Paul Du Chaillu as a *makaga*, an official responsible for enforcing laws and punishing criminals.[12] The aging ruler also granted other American and English traders the right to lease land and work as agents of his authority between 1859 and the 1870s.

Such arrangements never constituted a surrender of formal sovereignty and reaffirmed the rights of Nkomi leaders to control foreigners. Even when the arrival of a French gunboat in the lagoon in 1881 forced the ruler to sign away his independence again, Oyembo Onanga still contended he had the right to control Westerners in his territory. After his demise, individual clan leaders controlled sections of the lagoon over the course of the late 1880s and 1890s. Garner arrived just as the old order of Oyembo Onanga had shattered into many small principalities. Fernan Vaz was far from Libreville and the Ogooué River, the main means of communication in Gabon, and thus became a low priority for the colonial administration. French officials did little to enforce their will in the region besides occasional visits by gunboats to support a handful of officials scattered along the Fernan Vaz and Sette Cama Lagoons in the late 1880s and 1890s. As late as 1897, Nkomi leaders angry about high prices at English and German trading houses forced all European merchants to evacuate the region.[13] Garner had managed to come to Fernan Vaz just at the moment when central authority had collapsed, but before the French government had clearly asserted its authority.

From 1900 to 1902, Garner lived north of Fernan Vaz near a village occupied by members of an Orungu Omyènè–speaking clan, although hardly any documentation exists about his stay there. Between the mid-eighteenth century and the 1870s, Orungu clans had become among the wealthiest communities in all of Gabon, thanks to their pivotal role in controlling European access to the Ogooué River, which served as the main means by which slaves and natural resources reached the Atlantic coast from the Gabonese interior.[14] The French government and European trading firms broke the Orungu monopoly over access to the Ogooué by 1876. The consequent loss of trade left Orungu families few options other than selling fish and produce to the small French settlement at Mandji Island, just off the Gabonese mainland. By the turn of the twentieth century, Orungu people lived in widely scattered and politically divided towns. Garner may have preferred to stay near Orungu communities because their fragmentation made it very difficult for these villages to demand much from foreigners living nearby.

Farther inland, Garner encountered the two largest linguistic communities in the entire region: the Fang- and Gisir-speaking clans. Gisir speakers had long lived in south-central Gabon, and Fang clans arrived in Ngovè and Fernan Vaz only in the 1880s, but they shared much in common.[15] Aspiring men sought to trade ivory, rubber, and other natural resources for imported merchandise. They then exchanged their profits to other families in return for the right to marry women. Control over women's labor and having many children constituted for Fang and Gisir men the greatest sign of wealth. Women themselves often were forced into marriages but could assert their interests through abandoning husbands or joining female spiritual associations linked to healing (*ombwiri* in Gisir, *mekoma* in Fang). Divided into several dozen clans, Fang and Gisir politics were extremely decentralized, but male spiritual associations (*ngil* in Fang, *mwiri* and *bwiti* in Gisir) allowed for interclan alliances and offered protection from spiritual forces. Perhaps the largest distinction between the two came in their participation in the slave trade. Domestic slavery did not exist in Fang communities, but Gisir men brought captives from farther in the interior down to the coast. Nkomi communities maintained friendly relationships with Gisir clans, in part because of their mutual participation in the slave trade, but generally viewed the Fang as upstarts that threatened trade and order in southern Gabon

Along with the Nkomi clans that had dominated coastal trade on the Fernan Vaz and Ngovè Lagoons, two smaller communities also resided in the region. It is unlikely that Ngovè or Varama people comprised more than twenty thousand people in the early twentieth century, but their reputations for fishing and arcane supernatural knowledge gave them leverage in their dealings with their more numerous Fang, Gisir, and Nkomi

neighbors.[16] Ngovè leaders tried also to build partnerships with visiting
Europeans. One Ngovè leader even tried to place the entire Ngovè Lagoon
under British authority in the mid-1880s, because English traders were
more popular than French authorities. The Ngovè village chief Èdembè
y'Igowè, who died around 1920 reputedly at the age of one hundred, employed his knowledge of English and his claims to have mastered powerful
water spirits residing in his lagoon to ensure his people were not driven
off by Fang migrants. Garner, Charles Aschemeier, and other Americans
visiting Ngovè all turned to Èdembè to learn how to hunt gorillas.[17] The
Varama people generally lived south of the Fernan Vaz and Rembo Nkomi
regions, but some clans had established hunting and fishing settlements.
Although very little has been written on Varama communities, their ability
to command mystical forces remains respected among others living along
the Rembo Nkomi in the early twenty-first century.

Gender profoundly shaped everyday life in these clan settlements, regardless of linguistic differences. Save for the preparation of fields during the summer dry season, women performed most agricultural tasks:
planting the staple foods of manioc, plantains, and sweet potatoes, weed-
ing fields, harvesting crops, and pounding manioc tubers and plantains.[18]
Because most fields were far from villages, women bore heavy loads of
firewood and produce in large baskets on their backs. Free men frequently
traveled far from their settlements to collect rubber from latex vines, an
arduous task that required men to cut vines, collect the sap, shape the
liquid into balls, and bring them back to their homes for eventual sale.
Others hunted animals, cut ebony logs, raided other communities, or tried
to find work with Europeans at Cape Lopez and trading houses.

However, it would be a mistake to assume that women remained merely
passive victims of male relatives and authorities, even if the vast majority of Garner's dealings with Gabonese people involved men rather than
women. In 1905 Garner blamed the death of his Nkomi friend Anjanga
on one of his wives and several slaves. After Anjanga's death, his mother
Oliga contended that she was Garner's new host and patron.[19] Garner's
erstwhile guest Ida Vera Simonton recalled meeting an elderly woman
who had once traded independently and owned many slaves before the
establishment of concessionary companies. Women of less prominence defied relatives by leaving husbands or seeking out support from missionaries. To the horror of Simonton, one Gabonese woman left her husband

for a French lover. Once the abandoned man married another woman, his former spouse returned to his village and demanded that the new wife immediately leave the man's home. If she did not comply, then colonial guards would come to beat the cuckolded husband at the behest of the unfaithful wife's lover.[20] Such cases reveal how women could act to uphold their own interests, even if female Gabonese rarely received much attention in Garner's accounts.

Slaves, like free women, also occasionally enter Garner's narratives, although they rarely surface as individuals. The onset of French rule in the late nineteenth century did not lead to the end of domestic slavery in southern Gabon. Even though the region played an important role in the Atlantic slave trade in the first half of the nineteenth century, it is difficult to uncover how slaves and free people negotiated with each other in southern Gabon in the late nineteenth and early twentieth century. Hardly any historical works treat the topic of the gradual decline of Gabonese slavery under colonial rule. Officials chose a very different approach in dealing with slavery than they did in French West Africa, where administrators formally banned servitude in 1905. Simply put, administrators simply ignored slavery altogether and did little to either promote emancipation or support the demands of free people. Slaves occasionally appear in Garner's writings and little suggests that slavery was in decline while he lived in Fernan Vaz. Garner's service as a soldier in the Confederate army and his opposition to abolition in the United States might help explain his unsympathetic opinion of Gabonese slaves. He accused some slaves of poisoning his close friend Anjanga in 1905, and others made off with his canoes.[21]

Although Garner never identified his sources regarding slaves, it is likely his prejudices formed in his own childhood in antebellum Virginia corresponded with the attitudes of free Gisir and Nkomi people. Free people mocked slaves for their poverty and their inability to speak Omyènè well, accused them of using supernatural forces to kill free people, and often left elderly slaves to die outside of their villages or cast their corpses into rivers without funerals. However, Garner could well understand how free people esteemed the esoteric knowledge of some slaves. Free Nkomi people respected the male power association of *bwiti* brought by slaves from Mitsogo clans, because *bwiti* leaders supposedly could find sorcerers through their ceremonies.[22] Several slaves had earned a reputation as healers in Abingdon shortly before the Civil War, and a very young Garner had even purchased a concoction from one of them.[23] Just as free southern Gabonese people and Garner could find common ground on slavery, the American's early life as a Confederate soldier and his resentment over the Reconstruction era influenced how he viewed the establishment of colonial rule over southern Gabon.

The Ambiguous Beginnings of Colonial Rule in Southern Gabon, 1885–1919

The death of Oyembo Onanga and collapse of the monarchy opened opportunities for individual foreigners to establish their own domains, even if the French government claimed to control the entire Fernan Vaz Lagoon after the king was forced to sign away formal sovereignty in an 1881 treaty. The limited resources of the French administration precluded officials from clearly supplanting Nkomi, Gisir, and Fang leaders as the sole authorities in the region. Expeditions against political leaders farther in the interior took precedence over those on the southern Gabonese coast. Occasional visits of French gunboats supplemented a tiny group of guards and administrators in the 1890s, but the quality of these authorities left much to be desired. Albert Veistroffer, a former associate of French explorer Pierre Savorgnan de Brazza, who commanded the region from 1897 to 1900, wrote in despair of the alcoholics, sadists, and chronically ill men who had staffed the district in the 1890s.[24] Veistroffer's immediate predecessor, Auguste Forêt, claimed to be a Nkomi king in his own right, married multiple Nkomi women after the death of his French spouse from fever, offered to pay a fine imposed by the government on Nkomi villages that had threatened British traders, and returned runaway slaves to their Nkomi masters.[25] Although little documentation exists about Garner's stays in Fernan Vaz for the 1890s, such chaotic conditions must have allowed Garner a great deal of independence in his own right. For three years Veistroffer himself struggled to leave "such a forgotten place."[26]

Compared to the government, Catholic missionaries were much more influential as patrons to Nkomi people. After the foundation of Sainte Anne in 1887, Father Joachim Bichet established a set of reciprocal relationships with Nkomi clans.[27] Some clan leaders invented a title for Bichet, the *renima*, and declared he was an independent political authority on the lagoon. To acquire students for his mission, Bichet received the right to "marry" girls. His spouses stayed at the mission school until they reached adolescence, when they were married to male converts. Missionaries viewed this maneuver as a sign of fealty by Nkomi vassals, but Nkomi people a century later presented Bichet as an equal to any Nkomi leader. Madeline Yeno, an octogenarian in 2009, told the story of when her mother, Elisabeth, began to attend the mission school. When Bichet's canoe arrived in the mid-1890s, nearly the entire village fled out of fear that Bichet was a Portuguese slaver. Elisabeth's grandmother remained, and she waved a machete at the priest to drive him off. Only Bichet's copious offerings of trade merchandise appeased her anger and obtained her blessing for the child to go to school.[28] Nkomi leaders sometimes menaced missionaries with boycotts as well, but the patronage as well as the priests' ability to mediate with the government ensured an uneasy alliance.

A new organization linked to the colonial government radically altered this complicated set of relationships, when in 1899 the French administration granted the Compagnie Commerciale du Fernan Vaz (CCFV) a monopoly over all trade in the Fernan Vaz region. Meanwhile, the Société du Haut-Ogooué (SHO) received over a quarter of the entire colony as its own fiefdom. The government's miniscule budget and the dominance of English and German trading firms in its economy made Gabon an embarrassment in the French colonial bureaucracy. In the late 1890s Pierre Savorgnan de Brazza, the architect of French imperialist expansion in central Africa, decided to grant monopolies over trade to concessionary companies willing to promote economic development.[29] This program proved to be nearly as catastrophic in southern Gabon as it had been under Leopold II of Belgium's notorious Independent State of the Congo. English and German traders could operate only in the small section of the northern Gabonese coast that included Cape Lopez and Libreville. The low prices and poor selection of merchandise offered by concessionary companies angered many Gabonese people, as did the French government's constant efforts to collect taxes in French francs. Because concessionary companies preferred to overcharge Gabonese customers by trading only in merchandise and constituted practically the only source of European currency for rural Gabonese, many villages lacked the means to pay the annual poll tax instituted in 1902.

The CCFV's economic monopoly in the Fernan Vaz region and the collection of the poll tax brought on an economic crisis of catastrophic proportions in southern Gabon. Famines became a serious problem from 1905 to 1907, when so many families fled from colonial guards seeking to collect taxes and enforced laborers that food production decreased significantly.[30] CCFV agents offered Fang, Gisir, and Nkomi men a fraction of the merchandise and money that English and German traders had paid for rubber and ivory before 1899. Political leaders bemoaned how dramatically prices had dropped. A Fang clan chief from the Sette Cama Lagoon compared the situation to an unhappy marriage in 1907: "A Fang man had three wives who competed with one another in their love for their husband. One of them died, but the two other wives continued to serve their husband with the same zeal. However, when still another wife died, the husband began to endure the worst tribulations. The surviving woman became so arrogant that the husband chose to divorce her. . . . It is the same situation with the CCFV, since it no longer has any rivals."[31] Gisir and Nkomi men also complained about the CCFV in similar fashion, but the governor of Gabon's office refused to intervene, despite the support southern Gabonese complaints received from some low-ranking administrators.

In 2007 and 2008, over twenty informants in central and southern Gabon told stories about the hardships of early colonial rule. Nze Ndong, an elderly woman born in the Fernan Vaz town of Kongo around 1910 declared a guard had kidnapped her mother and forced her to move to the coast.[32] Villagers hiding from guards assigned to collect taxes cut the throats of their chickens to ensure the birds did not give them away. Rural people from Cocobeach in northern Gabon to Fernan Vaz accused guards of raping local women. Some villagers fled from the banks of the Ogooué River to forested hills so they could quickly conceal themselves at the approach of guards.[33]

Incessant demands for taxes compounded by the economic crisis created a crisis of political authority. Southern Gabonese political leaders had relied on access to imported goods to build alliances and obtain dependents, whether slaves or wives. French officials appointed some Africans to serve as chiefs to govern various districts, with no concern for differences between clans. Olago Vandji, a man born to a German father and a Nkomi mother, became the official chief of the entire Fernan Vaz region in the first decade of the twentieth century.[34] While he profited from state patronage, other clan leaders struggled just to pay their taxes. Richard Garner's close friend Anjanga, leader of the Nkomi Arunduma clan, kept asking Garner for enough money to cover his tax burdens in 1904 and 1905. The free trade zone of Cape Lopez became a popular destination for southern Gabonese people fleeing the concessionary companies. Still others, particularly Gisir clans, withdrew into remote forested areas to avoid paying taxes.

Gabonese people displayed their anger in different ways, but no major armed rebellions occurred on the southern Gabonese coast. The forested mountains of southern Gabon, populated by Mitsogo-speaking clans famed for their mastery of esoteric supernatural powers, was the site of one of the fiercest guerilla wars between the French military and Gabonese rebels between 1904 and 1920. Other than sporadic unrest by individual Fang clans, few Gabonese on the southern coast dared test directly the might of the colonial government. Gunboats carrying Senegalese soldiers could easily reach and burn down most settlements. The French government's decision to ban the sale of repeating firearms to Africans and to restrict the importation of gunpowder made armed resistance increasingly difficult, especially after the colonial administration signed an international agreement in 1908 to severely curtail shipments of firearms and gunpowder into the colony. French officials praised the relative calm of the region as proof of their authority, but the lack of violence on the part of Gabonese people testified more to their pragmatism in the face of overwhelming military force instead of their support for colonial rule.

Gabonese people scrambled to find support against the government and the CCFV. Missionaries ultimately had little influence on the CCFV or the colonial government, and local people sought out other foreign supporters for their own interests. In contrast to the concessionary companies, English-speaking traders received great praise from Fang, Gisir, and Nkomi men. American, British, and German traders all communicated with coastal Gabonese people in pidgin English, and all became known as "the English." Nkomi clan leader Olago Vandji informed a colonial inspector in 1906, "The country has been plunged into misery since the English left."[35] A year later, Fang and Nkomi men told Garner's companion Ida Vera Simonton again and again that they wanted the English to return to their region because they had paid market prices for natural resources in high quality merchandise. The angry man whose wife had abandoned him for a Frenchman and whose harp was confiscated by an administrator, said in pidgin English, as he pointed to the old, dirty cloth he wore, "I be proper frien' for Amerike woman, but French he all time make thief palaver for black man. My part, got no more. Every damn thing thief palaver. Them harp, him be my woman's. Him no be my part. My woman her live for white man for factory. He no be my part no more. White man thief her."[36]

Despite Garner's virulent belief in the fundamental differences between black and white people and his dislike for all things European, he became associated with the lost prosperity of English-dominated Atlantic trade. His American background had particular resonance for Nkomi people. First, southern Gabonese people had preferred Virginia tobacco to the varieties sold by the CCFV.[37] By the late nineteenth century, smoking had become so popular that tobacco leaves were one of the most commonly used forms of currency in the entire colony.

Second, although no American Protestant missionaries had ever set up a station in southern Gabon, despite the 1842 foundation of the Baraka mission in Libreville, one American had also lived in Fernan Vaz well before Garner: the American sea captain Richard Lawlin. From 1859 until his death in 1861, Lawlin ran a trading house and had been the most popular foreigner trader to live in Fernan Vaz. Nkomi men happily recalled to the English traveler Winwood Reade in 1863 how Lawlin had become an official serving Oyembo Onanga and had advised the monarch on legal policy.[38] Simonton also heard of Lawlin's fabled generosity: "The Nkomi denied that their land belonged to the French, but instead said their country belonged to Lawler [sic]. He bought it with 'proper rum, proper cloth, plenty powder, and proper guns.'"[39] Garner also heard similar stories about his compatriot. A family still held a tattered American flag and the remnants of a military uniform that had belonged to Lawlin, and the island where he had lived was known as Brooklyn. "It also appears that Lawler [sic] dictated certain laws which are still in force and certain trade regulations that he established are still today conformed to," Garner wrote.[40]

Few details have survived about Lawlin's life, but his historical memory served as a foil to the claims of French commercial and political dominance in the early twentieth century. Lawlin had accepted Nkomi rights over land, acted as a generous source of imported merchandise, and provided advice rather than expecting blind obedience to his commands. In the second half of the nineteenth century, English traders sometimes quarreled with Nkomi property owners and neighbors over prices and the right to trade freely on the lagoon, but local people likewise imagined their former commercial partners as munificent in comparison to the stinginess of the CCFV. Garner's national identity thus benefited his relationships with southern Gabonese people, especially as his American background linked him to an earlier period of Atlantic trade in which Africans retained the upper hand in negotiations. Instead of being a Christian opponent of local spiritual beliefs or a colonial official who enforced policies that left southern Gabonese people impoverished, Garner could present himself as a kindhearted patron who supported free trade and respected local political leaders, as Lawlin had.

However, southern Gabonese people had more practical reasons to support Garner. The French government allowed Garner to freely hunt and purchase animals, and so he was the only independent trader in the domains of the CCFV. French administrators did not develop a coherent policy regarding conservation in Gabon before the end of World War I, unlike the strict regulations on land use in French Algeria. Garner urged his patron William Hornaday, the director of the New York Zoological Society, to encourage French conservation groups to promote new hunting regulations in 1913 to "arrest the reckless and murderous traffic" in chimpanzees and gorillas. Despite continually seeking out both live and dead animals himself, Garner lamented to Hornaday, "I have already given you some rough statistics of the vast numbers of these creatures that are annually sacrificed to the greed of mercenaries."[41] Hunters, porters, fishers, carpenters, and canoe workers all worked for Garner. Some men, such as the Fang hunter Djego Ndong, may have worked for Garner for almost two decades.[42] A series of men also cooked for the American, although Garner's cantankerous treatment of his domestic servants must have not made him popular.

Garner's cooks and servants belonged to a diverse coterie of Africans that traveled on the Atlantic commercial and political networks throughout Central and West Africa. The same steamers that brought together Garner's ambitions with Gabon carried thousands of men and women along the African Atlantic coast in the late nineteenth and early twentieth century. The establishment of the German colony of Cameroon, French

territories in central Africa, and the Independent State of the Congo all
created demand for skilled laborers in the 1870s and 1880s. English traders
employed by the Hatton and Cookson firm and the John Holt company
recruited bakers, carpenters, cooks, and masons from the British colonies
of the Gold Coast (Ghana) and Sierra Leone. When the French government banned English from mission schools in Gabon in 1883, the decision exacerbated the shortage of commercial agents able to speak fluent
English. French officials and trading firms preferred to recruit soldiers and
traders from Senegal, particularly the coastal enclaves of Gorée Island and
Saint Louis that had been bastions of French influence since the seventeenth century. Senegalese guards and commercial agents had a terrible
reputation for brutality and unscrupulousness throughout Gabon, but
some Senegalese men married with Gisir and Nkomi women. Gabonese
Catholic and Protestant mission graduates found work in the interior of
French West Africa, as well as in their home colony. No matter how much
Garner mocked his supporting cast for their supposed laziness and stupidity, he needed them to do business. European companies and governments
also could not have operated effectively in central Africa without migra-
tion labor.

The grim economic conditions of southern Gabon improved briefly between 1911 and 1914. Governor general of French Equatorial Africa Émile
Gentil chose to cautiously reopen trade in Fernan Vaz to other companies
during this period. German demand for *okoumé* wood, from a towering
tree whose flexibility and light density made it popular for its use as ply-
wood, led to high prices. English traders and the French Societe Agricole,
Forestiere et Industrielle pour l'Afrique (SAFIA) consortium began to purchase okoumé from local men. Because the French government did not
require authorization for tree cutting, this business was popular with the
Gabonese community. Fang and Gisir men moved from inland regions
to toil on the coast. Although this commerce dominated the southern
Gabonese coast from the 1920s to the 1960s, the start of hostilities between France and Germany in 1914 shut down this promising source of
income.

World War I violently shook the global circuits of colonial authority
and commercialization that Garner and his domestic servants had relied
on for so long. Garner left Gabon in the spring of 1914 with several chimpanzees and gorillas, and the rush of people fleeing the war forced Garner

to sleep on deck on a ship from Amsterdam to New York.[43] When he finally returned to Port-Gentil and Fernan Vaz in early 1917, the conflict had devastated Gabon for nearly four years. French and German armies clashed in northern Gabon and southern Cameroon from August 1914 until the fall of 1916. Thousands of Gabonese men carried weapons and supplies from Libreville on foot to the front. French administrators drafted Gabonese men to join the army, although the colony contributed very few troops to fight on the Western front.

Southern Gabon was far removed from combat. Yet the war did not spare its inhabitants from suffering. International trade with the region ground to a halt in the first month of fighting.[44] Germany purchased the majority of Gabon's growing timber exports, and U-boats successfully limited trade to the colony. The French government doubled taxes to support the war effort in August 1914 and then coerced Gabonese people to collect rubber for use in France. Heavy rains in the dry season of 1916 compounded these misfortunes, as did the brutal escapades of a French official named Charmanade, whose name became synonymous with cruelty until he finally was removed from office.[45] In desperation to find food, inland Gabonese communities turned on one another: neighbors stole from each other's fields, elderly men begged children for food, and the colonial government forced farmers from the southern Gabonese interior to move to Fernan Vaz to set up fields.

Garner and his young companion Charles Aschemeier found themselves struggling with the same difficulties as their southern Gabonese neighbors, even though their affiliation with the Smithsonian Institution and the French government's support for the United States provided benefits unavailable to Africans. "You people at home don't know what it is like to live in an African community where there is practically no food," Aschemeier wrote his superiors at the Smithsonian in October 1918.[46] Garner recognized the causes of the famine: a shortage of gunpowder that made fields vulnerable to attacks by animals, a very wet dry season, and the skyrocketing prices for all imported goods.[47] On a hunting trip through the Ntchonga plains on the shore of Fernan Vaz, Garner had to send workers sixty miles away to find plantains, normally one of the most common staples in southern Gabonese cuisine. Medicine and supplies became increasingly hard to acquire, and even finding an opportunity to travel from Fernan Vaz and Port-Gentil became difficult. Of course, hardly any southern Gabonese could afford imported goods or could count on the largesse of colonial officials.

During these troubles, mission-educated Gabonese called for reforms. Garner detested literate Gabonese people as signs of cultural degeneration wrought by missionaries, and so it comes as little surprise that he never discussed their efforts to overcome the racist prejudices and heavy-handed policies of the colonial government. Libreville and Port-Gentil became the main centers of unrest shortly before World War I. Jean-Baptiste Ndende, a former naval carpenter, formed a chapter of the French human rights organization *Ligue des droits de l'Homme et du citoyen* (LDH) in 1918.[48] Although the Gabonese LDH's demands for higher salaries for African clerks and equal rights for Western-educated men reflected its elite membership, its ability to mobilize connections in the French metropolitan government placed officials at a major disadvantage. Prior to the LDH, complaints by Gabonese people had to go through the local administration to reach the minister of colonies in Paris, unless individual officials or missionaries chose to intervene. Between 1918 and 1927, Ndende and his colleagues sent a flood of telegrams to the LDH central committee in Paris that criticized high food prices, plans to physically segregate Gabonese towns, and the brutality of guards in southern Gabon. While Garner contended that Gabonese people lacked the intelligence to defend their own interests against the depredations of French companies and the colonial state, the LDH demonstrated the effectiveness of coastal Gabonese men to assert themselves.

When Garner last left Fernan Vaz in March 1919, southern Gabon had reached the end of a very turbulent era. Only Mitsogo clans in their mountain strongholds managed to retain a semblance of autonomy through the 1920s. Fang, Nkomi, and Gisir communities had lost their independence. The French government rescinded the rights of the CCFV by the end of the war, and dozens of French citizens moved to the southern Gabonese coast to establish timber camps. Thousands of migrant workers descended on the coastal lagoons, thanks to the resurgence of the timber industry. The need to save enough money and merchandise to pay dowry drew these workers to the lagoon, as did the hated poll tax. In this new era, colonial administrators continued to ensure that Gabonese people would have little control over the riches that came from exploiting the region's environmental resources. A system of expensive permits made it very difficult for all but the wealthiest Africans to run their own timber businesses. Over the course of the 1920s and 1930s, the French government restricted the hunting of big game, in addition to the already tight controls over the purchase and ownership of firearms. Although urban Gabonese men formed political associations that forced the colonial government to agree to some reforms, rural communities in southern Gabon had little political influence. Garner thus had witnessed the success of the colonial state in subjugating southern Gabonese communities and stripping them of control over the natural resources in their domains.

Oil became the most lucrative export of the southern Gabonese coast over the next four decades.[49] The colonial state and its postcolonial successor ensured yet again that indigenous communities never touched more than a tiny fraction of the profits. Less than twenty years after Garner's final farewell to Gabon, American and French oil companies began to search for petroleum in the region. By the late 1950s, the first shipments of oil began to make their way from the southern Gabonese coast to Europe and the United States. After Gabon gained its independence in 1960, the authoritarian regimes of Léon Mba (1902–1967) and Omar Bongo (1935–2009) worked closely with the multinational giants Elf-Total and Shell. Profits from this arrangement allowed high-ranking officials and their families to buy French châteaus and expensive sports cars, especially after the global rise in oil prices after 1973. The growth of Port-Gentil as Gabon's economic center allowed farmers and fishers on the southern coast a well-paying market for their produce. Some Gabonese men made high wages serving on oil rigs, but the vast majority of petrodollars never made their way back to Fernan Vaz, Ngovè, or Sette Cama. Most of the residents of Ombouè, Garner's home base during World War I and now capital of the Etimbwe district, live today in the same style of tin-roofed wooden-plank houses that most Gabonese call home.

The beauty of southern Gabon has inspired a new form of concessionary company, established almost a century after Garner's first visit to the region. Operation Loango, a company established by several wealthy European expatriates, received a grant to promote ecotourism through the Ngovè and Sette Cama Lagoons.[50] Wealthy expatriates are invited to enjoy pristine beaches, tour rainforests to see elephants and gorillas, and stay in lodges that cost several hundred euros per night. In the early 1990s the firm even occupied the Sainte Anne mission, as the declining numbers of French missionary priests left the church buildings vacant for several years. President Omar Bongo allowed Operation Loango to administer a new national park that he established in 2002. Residents of the small towns of Assewe, Sainte Anne, and Kongo on Fernan Vaz accused Operation Loango of running roughshod over local land rights in late 2008 and early 2009. Some expatriates working on environmental issues share the same complaints. However, the Gabonese national government's commitment to ecotourism makes it extremely unlikely that Operation Loango will be seriously threatened by local people opposed to its policies. Foreign consumers and companies continue to profit from southern Gabon's natural advantages, with comparatively little benefits for the inhabitants of the region.

Comparative Perspectives

For the scant number of Americans and Europeans who became interested in Gabon in the early twentieth century, they immediately compared it to another colony that had drawn far more attention: Leopold II of Belgium's Independent State of the Congo. The scandal of colonial oppression in the Congo was easily the most common article of interest about central Africa in the late nineteenth- and early twentieth-century Anglo-American press.[51] Concessionary firms received gigantic fiefdoms, in part to raise revenue to pay for colonial occupation. Leopold II ruled over the colony directly as a state entirely separate from the Belgian government. Investors and the colonial regime thus shared a common need for revenue and could not count on the Belgian parliament to provide funds for the colony on a regular basis. Growing demand for rubber drove the colony's economy, and government officials and company agents brutally exploited Africans to ensure rubber exports remained high. A diverse coalition of Belgian socialists, Protestant missionaries, evangelical Christians in England and the United States, and intellectuals such as Mark Twain protested against the violence and exploitation in the Congo. British journalist and politician Edmund D. Morel formed the Congo Reform Association to demand systematic changes in the independent state. Protests to European governments, journalistic accounts, rallies, and petitions eventually created so much negative publicity that some English politicians advocated the confiscation of the Congo and its redistribution among other European states. In a typical display of cynicism, Leopold II finally agreed to end the crisis by selling the colony to Belgium in 1908.

Morel contended that the iron bond between concessionary companies and the French government in Gabon was nearly as horrible as in the Independent State of the Congo. Some leading figures in France concurred. Pierre Savorgnan de Brazza, the Franco-Italian explorer who had led French efforts to control much of central Africa, agreed to return to Gabon and other French colonies to investigate allegations of abuses by concessionary companies.[52] While he found much evidence to support the charges, his death in 1905 en route to France furnished an excellent opportunity for the colonial ministry to effectively halt the inquiry. Despite the polemics of some intellectuals, such as the journalist and human rights advocate Felicien Challaye, Gabon and other French territories in central Africa did not draw much interest in Europe or North America.

However, another reason may better explain why so few Americans and Europeans paid attention to the effects of colonial rule in Gabon: publicity. French Catholic missionaries knew full well how badly many Gabonese were faring under colonial rule, but very few of them chose to expose their findings to the French press. Dozens of English and North American Protestant missionaries in the Independent State of the Congo provided photographs and testimony of the systematic maiming of children and adults. Protestant missionaries in Gabon, by contrast, were largely French Calvinists, and their missions were far from the concessionary companies' domains. American Congregationalists and Presbyterians had established churches in the colony since 1842 but chose to concentrate their efforts on southern Cameroon by the 1890s. By 1910 the Presbyterian West African Mission chose to surrender their remaining churches to the French Société des Missions Evangéliques de Paris. The colony's relative obscurity did not draw many visitors, outside of a few animal specimen collectors and wealthy world travelers. Garner was among the tiny number of American and European travelers who spent much time in Gabon. Naturalist Mary Kingsley was the most well-known writer to come to Gabon, and indeed she and Garner both were in Gabon in 1895. However, she left before the concessionary system had actually been implemented. All in all, a lack of exposure helps explain why so few knew or cared about the violence of colonial occupation in Gabon.

For those familiar with the opening stages of the rapid European inva-
sion of African states in the age of high imperialism, the argument that
Appalachia also experienced a form of colonization in the late nineteenth
century seems far-fetched at first glance. Gabonese people could not vote,
had practically no legal recourse to complain or challenge decisions made
by state authorities, and had their land rights stripped away with a few pen
strokes made in Paris. Certainly, Native American communities underwent
a similar process in the mountain South. Although many men and women
in the mountain South rejected secession due to their antipathy toward
the dominant planter class, the Civil War and the violent disruptions of
the Reconstruction era also brought about difficult challenges. In the 1960s
and 1970s, a group of activists and scholars contended that northern commercial and political elites imposed their will over local Appalachian communities, in similar fashion to colonial governments elsewhere.[53] With the
coming of the coal industry after the Civil War, large mining firms exerted
their considerable influence to acquire vast land rights. Courts, along with
state and local governments, supported commercial interests against trade
unions and farmers.

Increased industrial expansion and the loss of local autonomy to national governments left in its wake both winners and losers in the late
nineteenth and early twentieth century, even if conditions widely varied
between countries and regions. Appalachian men risked their lives in coal
mines only to see the bulk of the profits go to company owners. Gabonese
families lost control of the timber, rubber, and ivory trades. Local governments became increasingly subservient to the interests of remote commercial interests in the southern United States and central Africa. Racist
ideals of French officials and white politicians and intellectuals considerably differed, but their perspectives reinforced a common belief in white
supremacy that disenfranchised people of African descent. Appalachian
families chose to leave behind the poor economic prospects of their mountain origins for new lives in Baltimore, Washington, Atlanta, and other
cities, just as southern Gabonese people headed for the small urban settlements of Port-Gentil and Libreville. The process of dispossession and
the radical remaking of local politics were quite different in Appalachia
and southern Gabon, but the results of endemic poverty had much
in common.

The binaries of resistance to and collaboration with this new economic and political order dominated the historiography of Appalachian and African history in the 1960s and 1970s. Scholars have grown dissatisfied with this Manichaean divide in the last three decades.[54] First, such a sharp split ignores the importance of class and gender divisions. What constituted losses for some Gabonese in this era opened opportunities for others. Slaves and women sometimes took advantage of the troubled state of local Gabonese political institutions in the initial decades of French colonization. African American communities in southern Appalachia strove to establish schools and assert their political rights in the Reconstruction era, even though white supporters of the Confederate cause portrayed Reconstruction as an unmitigated disaster.

Individuals in Appalachia and southern Gabon selectively opposed and appropriated ideas and practices from outside their home regions, rather than forming a united front against the commercial and political forces transforming their communities. Gabonese communities often both resisted and supported state policies, and the very popularity of unrestricted trade in rubber and ivory suggests that Gabonese traders happily entered the global economy, as long as they could decisively control the terms of trade. Nkomi and Gisir clans formed close ties with Catholic missionaries, even as they largely refused to abandon local spiritual traditions. Coastal Gabonese intellectuals mastered French republican rhetoric of civilization and equality to challenge French administrators.[55] In Appalachia, middle-class townspeople actively participated in the representation of their home region as primitive, unique, and separate from the rest of the United States. Townspeople also sought to distance themselves from other Appalachians through their depictions of rural farmers.[56] By highlighting the supposed ignorance and strangeness of southern rural mountain society, Appalachian writers presented themselves as a bridge between dominant middle-class values and the exotic otherness of their homeland. Coastal Gabonese fluent in English and French played a similar role by displaying their superiority over inland communities while acting as interpreters and intermediaries to officials, missionaries, and traders. Garner thus was an intermediary, relying on his Atlantic network of commercial and political connections to make a living, as were so many Gabonese people coping with the hardships of colonial rule.

Conclusion

Southern Gabonese coastal communities underwent dramatic and painful changes between Garner's initial visit in 1892 and his final departure in 1919. Older political and commercial networks built on clan alliances and partnerships came undone after 1899, thanks to French colonial guards, taxes, and concessionary firms. The collapse of the centralized Nkomi Kingdom created a power vacuum in which rival clan leaders and entrepreneurs sought to make their fortunes. Relying on brutal displays of force, French authorities demonstrated their ability to vanquish armed resistance. Yet the administration proved far less able to promote trade, and the concessionary company regime ensured the paralysis of the southern Gabonese economy until the reintroduction of foreign competition in 1911. World War I then battered the region, leaving many Gabonese people living on the brink of starvation. Richard Garner thus was a witness to the tumult of the initial decades of colonial rule in Gabon.

Garner's lengthy writings on Fernan Vaz offer a unique perspective on the ways Africans grappled with the weighty loads placed on them by concessionary companies and colonial governments in the early twentieth century. While he thoroughly rejected any similarities between his own ambitions to overcome his origins and the efforts of coastal Gabonese people to defend their own interests, the Appalachian adventurer and his Gabonese neighbors both relied on creative alliances to support themselves. Garner's position as an American not clearly connected to any of the major power brokers in the colony, and his willingness to recognize indigenous rights to land and unrestricted trade, allowed him to develop long-lasting relationships with Gabonese people. Garner obtained so much information on indigenous cultural practices because of his begrudging willingness to treat Gabonese people as equals, even though he considered Africans fundamentally inferior to whites. It is testimony to the desperate condition of Gabonese communities that even a cantankerous eccentric like Garner could be seen as preferable to colonial authorities and French company agents.

CHAPTER TWO

Garner's Animal Business in Africa and America

THROUGH THE WRITINGS OF RICHARD GARNER, this chapter brings together the heretofore distinct but intimately related histories of French colonialism, the African environment, and North American zoos. These texts provide a rare glimpse into the connections between colonial economic and social policies and animal shipments to zoos, and they also represent one of the most cohesive and complete collections available of work by and about an individual animal collector in west-central Africa. Between 1892 and 1919, Garner became an acclaimed supplier of chimpanzees, gorillas, and other exotic animals for the New York Zoological Society (NYZS) and the Smithsonian Institution. While he portrayed himself as a lone and intrepid hunter struggling against dangerous beasts and wild Gabonese people, he in fact depended on an extensive network of indigenous hunters and traders, French officers, and European merchants.

Although many Americans often think of zoos merely as a place to see and encounter animals, they were venues that both affirmed racial hierarchies and relied on colonial rule. Historians have long noted how the wealthy businessmen and politicians who sponsored the creation of new zoological gardens in the United States during the Gilded Age, most notably the National Zoo in Washington and the Bronx Zoo in New York City, hoped their new establishments would serve to reaffirm Anglo-Saxon superiority over other races.[1] Contact with wild nature, no matter how carefully managed, would lead to white regeneration in an era when fears of "hyper-civilization" haunted many Americans.[2] Madison Grant, the prime mover behind the New York Zoological Society, closely linked the Bronx Zoo to a defense of native-born white civilization against immigrants and racial threats.[3] Zoos confirmed racial hierarchies and relied on colonial administrators in African colonies to permit the importation of rare animals. These sites introduced visitors to foreign colonial territories, upheld the superiority of white Americans through highlighting the supposed primitive and dangerous nature of African life, and left silent varied African understandings of animals. Furthermore, zoos also showed the power of officials in African colonies to control access to animals. In discussions of the connections between imperialism, race, and zoos, the negotiations of animal collectors with African colonial subjects and European officials have rarely drawn attention.[4] This negligence is particularly ironic given that animals would have been extremely difficult to acquire without the services and the authorization of European authorities.

However, information on most animal dealers who sold African animals to American zoos is hard to come by, at least for collectors after the early nineteenth century. German zoo impresario and animal dealer Carl Hagenbeck has certainly drawn attention as an international animal supplier.[5] Yet we know precious little about other collectors, particularly the men who actually obtained specimens for Hagenbeck and other major players in the exotic animal market. Nancy Jacobs's recent treatment of George Bates, an American ornithologist who spent decades in Cameroon, is a rare exception to this trend.[6] Part of the problem lies in the nature of the sources; most collectors supplied a range of museums and zoos in the Americas and Europe, which makes tracking down these materials difficult. Animal collectors appear only fitfully in colonial administrative correspondence. The lack of environmental regulation in French Equatorial Africa before the 1930s, along with the disappearance of ephemeral records on gun permits and hunting authorizations, further obscures how animal suppliers interacted with Africans and colonial governments. Another complication lies in the orientation of most historical research on animal collecting and zoos, which remains squarely on Europe and North America. Historians of colonial Africa seldom consult museum and zoo archives. Scholars investigating zoos rarely engage in prolonged research in African countries or sift through state and missionary correspondence from colonial African locales.

The connections between American zoos and colonial governments deserve attention not only for what they reveal about American perspectives on race and imperialism but also because an investigation of animal collecting in west-central Africa for American markets is a crucial first step in the important task of recovering the environmental history of this region. Environmental histories are only just emerging on Cameroon, Equatorial Guinea, and all the countries that once belonged to the federation of French Equatorial Africa (Central African Republic, Chad, Congo-Brazzaville, and Gabon) in comparison to the rich literature on the environment in eastern and southern Africa. Likewise, the regulation of the environment in the French empire is just receiving attention now and is still easily dwarfed by the much more voluminous research on British imperial policies toward conservation.

Garner's correspondence denotes the varied activities of the agendas of African hunters, colonial officials, European trading companies, and rival animal collectors. Unlike in English and German colonies, the French government never attempted to create a coherent legal framework for regulating hunting until the 1920s and 1930s. Theoretically, officials concerned themselves only with enforcing increasingly strict policies on the purchase and ownership of firearms in the first two decades of the twentieth century. However, since the French government had granted a monopoly over all commerce to various concessionary firms in 1899, which remained in effect until the 1920s, these companies claimed that collectors such as Garner could purchase animals only with their permission. Garner alternately contested and accepted these arrangements as he sought out rare animals. At the same time, Gabonese hunters sought to acquire French francs from collectors to pay their taxes and to ensure that foreign and indigenous rivals did not learn of their own favored hunting grounds.

The tactics and strategies of animal collectors and hunters changed with efforts to reform the concessionary companies, and then again with the advent of World War I. Governor general of French Equatorial Africa Émile Gentil allowed some independent trading in concessionary domains after 1911, which helped Garner buy animals. Yet the simultaneous rise of *okoumé* timber exports led some of Garner's former African partners to abandon animal trading. Other developments further hampered animal trade. French administrators agreed to severely curtail gunpowder importation in Gabon after 1909 to follow international accords and to block supplies from reaching rebels fighting against the government. World War I increased the willingness of southern Gabonese people to trade animals with foreigners. Gabonese hunters sought out Garner as a rare source of currency and merchandise during the war, and the paucity of French officials allowed him to hunt without any impediment. The collapse of international trade to Gabon in the wake of the war made transporting and storing animals a very hazardous task. Very few ships visited the southern Gabonese coast, and even fewer were willing to take on animals or people. Live animals languished and dead specimens deteriorated in tin-roofed shacks, much to the consternation of North American zoo directors. Thus, collectors and hunters coped with changing commercial and political circumstances.

Richard Lynch Garner and Gabonese Animal Collecting, 1892–1910

Garner's presence in southern Gabon allowed him access to animals high in demand in New York, Washington, and other cities. Since Franco-American explorer Paul Du Chaillu had made the southern Gabonese coast famous in his descriptions of his search of gorillas in the 1850s, this region had attracted a range of hunters and tourists. Yet decades after Du Chaillu, chimpanzees and gorillas remained difficult to obtain even in this region. Although German scientific expeditions and travelers had regularly imported gorillas to Berlin, Hamburg, and London since the 1870s, the inability of these animals to survive the rigors of captivity and long ocean voyages made them extremely rare and fleeting sights in American and European zoos. Garner thus had found a niche for himself as an animal supplier in a region with an international reputation for primates, wild buffalo, elephants, and many less glamorous species of birds, monkeys, and other mammals. Sales of animals allowed him to maintain his research activities. As early as 1895, the Royal Ontario Museum had acquired the fur and skeletons of a gorilla and several chimpanzees from Garner.[7] The American zoologist depended on the profits from these transactions. With such a dubious reputation as a scientist, these sales became one of his central occupations. He boasted to his son in 1905, who had found potential investors willing to back Garner's acquisition business, "I am in touch with the zoos and taxidermists — and know where to find buyers for most of my stuff—so, you may assure your friend that his investment is not yet in any danger."[8]

Despite Garner's boasts about his heroic adventures, he relied on a loose network of southern Gabonese hunters to obtain rare animals for his American clients. While southern Gabonese communities had exchanged rubber and ivory freely with British and German traders since the 1860s, the decision of the French government to hand out commercial monopolies to a group of concessionary firms in 1899 brought a new era to Fernan Vaz.[9] Individual government inspectors noted how CCFV agents paid out to Gabonese hunters and workers desultory amounts of money and merchandise of poor quality and regularly beat their African employees. From 1899 to 1911 concessionary company agents claimed to have the sole right to export the natural products of Fernan Vaz. Even exotic animals were considered to be CCFV property. Peneau, a CCFV manager, sued a Gabonese man for selling a gorilla in 1906 to an English sea captain. A British trader, apparently stationed in the free trade zone of Cape Lopez, countered that he himself had bought the animal. Peneau may have acted to protect his own business, as he allegedly also traded gorillas to passing ships.[10] Fang, Gisir, and Nkomi men complained to Garner, missionaries, and officials about their declining fortunes, which they blamed squarely on the CCFV.[11]

How Garner managed to circumvent the CCFV's grip over trade remains unclear. Certainly, Gabonese men willingly traded with Garner. As an independent entrepreneur, the American paid clients and his workers hard currency and merchandise.[12] Some Gabonese seeking to be hired by Garner tried to flatter him. In a request for employment, one man praised Garner for his generosity: "Monsieur, you are a great man. Because you are a good man and also because you are an Englishman [*sic*], I ask you to give me work for 25 francs per month with daily rations included."[13] It helped that Garner went so far as to pay the hated poll tax, instituted in 1902, for some of his workers and supporters.[14] "The French government has been seizing all the chimpanzees and gorillas captured in [Fernan Vaz] for the native taxes and thus have forestalled me in several purchases—but I have a number of agreements with a number of [Fang] people. . . . I shall soon get what I want."[15]

Yet Garner's connections with many clans in southern Gabon did not constitute the sole explanation for his ability to do business in a concessionary company fiefdom. Given the tiny number of Europeans in southern Gabon, it would have been impossible to ship out animals to Cape Lopez and then the United States without the knowledge of the CCFV. More likely, French officials could have decided to grant Garner special permission to hunt and collect for scientific grounds, especially because he had collected animals for several years before the CCFV received its monopoly. Despite his frequent critiques of concessionary firms and the French government's taxes, Garner cultivated good ties with colonial administrators. Both Charles de Noufflard (governor of Gabon, 1906–1907) and Casimir Guyon (governor of Gabon, 1914–1916) maintained friendly relations with the American expatriate.[16] By 1909 Garner had convinced the CCFV to wholeheartedly support his efforts. Susie, the chimpanzee that rekindled Garner's fading popularity to audiences as august as President Taft and as large as the Bronx Zoo in 1910 and 1911, came from the director of the concessionary company. Garner praised the CCFV director in 1912, forgetting his virulent attacks on the company in previous years: The director "is the one who procured Susie for me and offers to assist me in any way. . . . He also offers me passage aboard the Company's launch . . . to Fernan Vaz."[17]

Garner depended on a network of hunters and traders that lived throughout Gabon. As early as 1899, a Fang man traveled several hundred miles from the Gabon Estuary to Fernan Vaz to tell Garner of a gorilla in his home village.[18] His diary entries in 1905 and 1906 report a steady stream of Fang, Gisir, and Nkomi men coming to sell bones, dead specimens, and live animals.[19] While Garner occasionally hunted on his own, his poor finances and declining health did not allow him to travel easily. Toward the end of his stay in Gabon from 1904 to 1909, Garner had to sell his only gun to pay for shipping animals to the United States. To acquire animals, he often had to travel inland to Gisir villages along the Rembo Nkomi River.[20] These hunters also recognized the shifting market value for animals, especially when multiple collectors arrived in Gabon. A doctor from the Pasteur Institute in Paris came to Gabon in 1904 to buy chimpanzees. Garner noted the results: The doctor "played the giddy goat all around, as he was spending the funds of a big institution. The result is that traders and natives have an idea there is a great boom in the ape market and everyone expects to realize a small fortune from one or two of them. . . . I am going back to the bush-forest belt of Esyira where I can get them from first hands and where the market is not so excited."[21]

The inordinate expense of shipping materials to the United States undid many of Garner's plans. A letter from 1900 reveals his incapacity to raise enough hard currency to pay for transport: "I have demonstrated . . . that I was able to go into the wild and make a collection of gorillas, skins, and skeletons that have not up until this time been equaled by any alleged zoologist. Yet I could not realize enough on them to pay the freights and I gave them away."[22] Sending out specimens involved a number of steps. First, Garner had to either go to where Gabonese people had captured animals or pay for their transport to his home in Fernan Vaz. Once in Garner's possession, the animals then needed to be brought by canoe or by the steamers that infrequently stopped at Fernan Vaz en route to Cape Lopez, the major port of entry into central and southern Gabon. There, arrangements needed to be made with English, French, and German steamers that regularly boarded timber, rubber, and ivory for European markets. No vessels went directly from Gabon to the United States, so specimens had to be transferred to ships headed for New York from European ports. Garner had little success ensuring that Gabonese animals survived these long and complicated voyages before 1912, unless he actually accompanied the creatures themselves to America.

Despite his questionable scholarly credentials and his high rate of failure, Garner eventually drew attention from North American museums and zoos. After his initial tour led the president of the University of Illinois, Selim Peabody, and some other Midwestern patrons to fund a second voyage in 1895 and 1896, funding appears to have dried up for the next decade and a half.[23] Few directors chose to take a chance on Garner before his tours with his pet chimpanzee Susie through much of the United States in 1910 and 1911. Garner bemoaned in 1900 how the Chicago Field Museum, the Smithsonian Institution, and other organizations wanted nothing to do with him.[24] Yet the tide finally turned in Garner's favor with the sensation of Susie, especially with famed naturalist and Bronx Zoo fixture William Hornaday. The two men shared a passionate belief in the need for conservation. Garner and Hornaday also espoused similar ideas on the biological fact of white supremacy.

Even with all of these similarities, their relationship did not begin well. The growing divide between amateur and academically trained scientists in the late nineteenth century made Garner's ascent a precarious one. It is unclear when Garner first began to reach out to Hornaday, but the director initially dismissed him. Hornaday displayed both his racist prejudices and his disdain for Garner in a children's magazine in 1894: "Recently a great stir has been made by Professor Garner, who has made the astounding discovery (?) that monkeys have a language of their own, and can talk to each other. Dear me! And who ever said they could n't [sic]? I suppose Mr. Garner will next discover that Africa is the darky continent, and the Dutch have taken Holland!"[25] Garner deplored Hornaday's lack of scientific credentials in a critique of one of the Bronx Zoo director's essays: "Of course Hornaday is only a taxidermist and it is not to be supposed he is up on mental science or linguistics—I can't imagine how he ever became the director of the bronx park zoo [sic]—for he is not a zoologist."[26] It seems striking that Garner attacked Hornaday over his training, because so many scientists lampooned Garner for his own position as an autodidact. Yet the two men eventually became partners in the primate trade.

The Garner Expeditions of 1911 and 1912–1914: Zoos, Collecting, and Colonial Transformations in Gabon

The New York Zoological Society chose to support Garner's two sojourns in Gabon immediately prior to World War I. His return to the colony came at a time of dramatic changes. After over a decade of economic decline, the southern Gabonese export economy began a short recovery that came to a grinding halt with the onset of World War I. Gentil also implemented a near-complete ban on the importation of gunpowder into Gabon, following France's decision to sign the Brussels accords that curtailed the introduction of firearms into African colonies. In Fernan Vaz colonial troops had forced rebellious Fang- and Gisir-speaking clans to accept French rule by 1911. Besides these policy changes, the sudden resurgence of *okoumé* lumber exports drew many Gabonese men out of hunting and rubber collecting into the timber business, whether as timber camp workers or as independent operators.

Although it is unclear when Hornaday chose to revaluate Garner's usefulness, there is no doubt about why the two men formed a close collaboration in 1911. The Bronx Zoo desired gorillas and chimpanzees. Garner wanted fame and money. Since 1905 the NYZS had tried in vain to procure a gorilla; every animal had died during the transatlantic voyage.[27] Garner's tours with Susie the chimpanzee had brought him wide acclaim, even though Susie's antics took precedence over Garner's theories of animal language. When Garner offered to sell Susie to the zoo in February 1911, he noted, "I have many offers to put her into vaudeville, but that would be to go into vaudeville myself and I am not a vaudeville attraction, I think, even if I wanted to go into that line, which I positively do not."[28] Hornaday eventually accepted the offer and within several months commissioned Garner to obtain a gorilla for the zoo, with a thousand dollars to cover expenses and a bonus of five hundred dollars for the successful delivery of a gorilla to New York.[29] The NYZS's powerful patron Madison Grant persuaded secretary of state Elihu Root to support Garner's travels to Gabon, which included letters of introduction from Root to the governor of the colony.[30] In search of adventure, Robert Whitney Imbrie, a young Yale graduate and attorney from a well-off Baltimore family, joined Garner on the voyage.[31]

After the small expedition arrived at Cape Lopez in May 1911, Garner found almost immediately that political and economic conditions had changed from his last departure two years before. Again, Garner operated as an intermediary rather than follow stereotypes of intrepid lone adventurers. "I do not find the prospects at all flattering for the rapid collection of gorillas because the native hunters can not get powder," he complained to Hornaday. Practically no powder could be obtained in Fernan Vaz; even colonial guards could not acquire it, thanks to a "comic opera act of legislation" that limited gunpowder sales to African colonies.[32] Even worse, "some fool came into Fernan Vaz a few months ago and bought a young chimp for which he paid 250 francs cash and since that time the natives had doubled, trebled [sic], quadrupled the prices of everything they have captured." To overcome these obstacles, Garner turned to his connections among officials and European traders. An administrator reported to the American acquaintance the unwelcome news that no gorillas had been captured in his district for over a year, but Garner felt secure that the government as a whole would support his efforts. After two months, the Americans managed to get a gorilla.[33] Yet again, he needed support from French residents and Gabonese informants. While Garner discreetly chose to not tell Hornaday how the gorilla came into his possession, he told journalists in New York that he convinced a recalcitrant French trader to surrender the animal after negotiating for a week. Gabonese men had first told Garner of the gorilla, and African hunters had initially sold the animal to the French merchant.[34] Even though the men had accomplished their goal, the hazards of transatlantic travel took their toll on their precious cargo, especially when some unforeseen delays forced Garner to make a stop in France. The gorilla already was ill when it finally reached New York in September 1911, and it died several weeks later.[35]

Undeterred, Garner persuaded Hornaday in the summer of 1912 to sponsor another Gabonese trip. By December the aging collector had already returned from New York to Cape Lopez.[36] He stayed in Gabon until the fateful summer of 1914, in part to prepare captured animals for diets based on bread to help the acclimation process. Another reason for his prolonged residence in Gabon came from problems in New York. When the entire collection of primates at the Bronx Zoo died from illnesses in April 1913, Hornaday asked Garner to buy more animals to replenish the zoo.[37]

Although French officials made Garner pay duties for bringing foreign ammunition into the colony, Garner's friends in the CCFV and the colonial bureaucracy went to work on his behalf. One officer offered to send Garner a gorilla by steamer from the southern Gabonese coast town of Mayumba. CCFV agents and other traders also volunteered their services and eventually allowed Garner to set up his camp for over a year on property owned by the company.[38] Because so few Europeans lived in Gabon—perhaps no more than five hundred or so before World War I—Garner's long residency in Fernan Vaz had allowed him to become an insider among the small number of officials and traders in the colony. To protect his interests against the financial demands of local officials and CCFV agents, Garner turned to the governor general of French Equatorial Africa and the company director.[39] It also helped that so many other Europeans already owned, hunted, and traded exotic animals, particularly monkeys and primates. It is a striking irony that Garner complained about the threat that unrestricted hunting posed to wildlife, because he profited from the free rein to hunt given to European residents of the colony. Admittedly, part of his criticism was aimed at his animal-collecting competitors, "the white shylocks who cheat the natives out of most of the commercial value of their [animal] captives," who threatened to wipe out apes and other species.[40]

Besides relying on his French colleagues, he also continued to depend on Gabonese workers. A team of eleven men erected Garner's home on the

Sette Cama Lagoon in the spring of 1913, and later Garner hired a dozen men to maintain his camp and to feed a gorilla he purchased.[41] Especially because Garner sought to train the gorilla to eat bread rather than only plantains, the cost of managing this staff and keeping supplies was fairly high. This work posed dangers; the gorilla reached through its cage and grabbed a Gabonese man by the hair. The American researcher also hired several hunters to seek out more animals.[42] Unfortunately, Garner did not preserve detailed records of his daily negotiations with his staff, so it is difficult to reconstruct how Gabonese workers obtained animals or why men chose to work for Garner.

Garner referred to several other Americans and Europeans who also sought out exotic animals. Charles Jesse Jones, a seventy-year-old former guide to Theodore Roosevelt and a retired game warden at Yellowstone National Park, decided to go with several associates from Wild West shows and a small film crew to Fernan Vaz in the spring of 1914.[43] Their goal was simple. Jones wanted to lasso a gorilla, just as he had roped various animals on a previous trip to Kenya in 1909. Apparently bankrolled by a circus, the expedition followed Hornaday's suggestion to go to southern Gabon. The Bronx Zoo curator wanted to combine Garner's fame with the celebrated "Buffalo" Jones to attract the American press, and Hornaday had induced the elderly man to give any captured gorillas to Garner, so that they would end up in New York.[44] Jones had already drawn fame for promoting the conservation of buffalo and did not promote the killing of animals. Jones's method of entrapping a live gorilla seemed utterly preposterous to Garner, and ultimately the old cowboy and his entourage had to settle for shooting two of the giant primates.[45]

Although the expedition failed to achieve its goal, Jones did describe his negotiations with Gabonese hunters and chiefs in more detail than Garner. Rather than seeking out Gabonese assistance, Jones met with a Ngove-speaking chief of some villages along the Iguèla Lagoon south of Fernan Vaz. This leader, Èdembè y'Igowè, approached Jones and told him how the colonial administration's policies had posed problems for his community: "He spoke broken English and told us he had heard of the white hunters far away, and had come two days to them [and] that in his country there were elephants, hippos, gorillas, and other wild beasts; that the French government had confiscated their guns and there was no way of killing them. He said the gorillas destroyed their bananas and polluted the drinking water." In the 1920s French adventurer Georges Trial heard similar complaints.[46] Later, Èdembè supplied Jones with guides and advice on how to hunt gorillas. He also demonstrated to the Americans how to prepare spear traps set in trees to kill elephants and how to set snares.[47] Yet again, the search for wild game by foreign visitors provided tangible benefits for rural Gabonese people and a means of circumventing colonial regulations. Èdembè would later reprise his role as an employee and a patron for Garner during World War I.

Garner faced more serious rivals than Jones, particularly a German animal supplier named Wedermeyer. This entrepreneur had contacted the Bronx Zoo and offered to sell Hornaday several gorillas from German Kamerun, directly north of French Gabon. Hornaday, desperate to bring as many apes as possible, instructed Garner to consider purchasing Wedermeyer's animals.[48] After receiving a photograph of one of Wedermeyer's gorillas, Garner accused his German competitor of exaggerating the animal's height by having two men hold its arms up. "All Mr. W's correspondence with me has a flavor of 'patent medicine'—He insists upon that 'only a few left' manner of his. . . . I don't mean to malign the man, but I know the 'gorilla stories' of the coast too well [to] be suddenly excited over them. . . . Don't allow yourself to be persuaded that Mr. W holds the 'last chance' in the gorilla market," he warned Hornaday.[49] It is easy to imagine that different animal dealers would seek to undercut the credibility of their competitors, and Garner's words succeeded in turning Hornaday against the German trader.[50] Garner must have been aware of the German Carl Hagenbeck, perhaps the most successful exotic animal supplier between 1880 and 1914, who sold the Bronx zoo several chimpanzees. Hagenbeck himself knew of Garner, at least enough to make jokes about how he could make monkeys talk as well as the famous American.[51] But Garner had proven himself against his German rivals.

After a year and a half, Garner chose to come back to the United States with his gorilla, several chimpanzees, and other specimens. Despite his extensive search through intermediaries throughout Gabon, he had learned of only four gorillas in captivity.[52] The onset of war made his return a difficult one. So many people had chosen to flee Europe that Garner had difficulty finding a room for himself and his gorilla, Dinah. Even worse for Garner and the Bronx Zoo, the news of Dinah's triumphant entrance into New York City on 21 August 1914 had to struggle against the bloodshed of the Western Front in American newspapers. However, Garner managed to find ways to bring attention to himself. Bronx Zoo assistant curator Raymond Ditmars recalled that Garner "cabled that he had found accommodations on a steamer, a small cabin for himself and a tool house on deck for the gorilla, but added: 'I am occupying the tool house and giving the cabin to the gorilla.' His humor and the comedy of the situation made such a hit that his arrival was written up in fair detail."[53] Hornaday praised Garner and noted how the war made obtaining more exotic animals practically impossible, especially since the British blockade ended the long reign of German companies over the zoo-supply business.[54]

Garner's work in Gabon for the Bronx Zoo relied on Gabonese hunters and his numerous contacts with the colonial administration and with European trading firms. Without Gabonese men eager to trade animals for hard currency to pay poll taxes in French francs, it would have been almost impossible to acquire gorillas. French administrators allowed Garner to buy animals and bring weapons into the colony despite a ban on gunpowder importation into French Equatorial Africa. His entire business thus depended on Gabonese expertise and the colonial government, instead of following the models of intrepid exploration. While the relaxation of the CCFV's grip over trade had allowed Garner more freedom to operate from 1911 to 1914, his ill-fated expedition to Fernan Vaz from 1916 to 1919 was a much more difficult enterprise. World War I unleashed a host of adversities on Gabonese people. Garner and his animals also experienced many of the same hardships.

Collecting Animals during a World War:
The Garner-Collins Expedition to Gabon, 1916–1919

Garner's ambitions only continued to climb after the success of his two expeditions supported by the New York Zoological Society, but ongoing hostilities made international travel a dangerous proposition. Meanwhile, southern Gabonese communities felt the impact of the war in ways far more serious than Garner's dilemmas.[55] The southern coast was spared the warfare between French and German colonial armies that wreaked havoc in northern Gabon, but this would be cold comfort to most Gabonese people living on the Fernan Vaz, Ngovè, and Sette Cama Lagoons. As Garner learned all too well in 1917 and 1918, the Gabonese export economy collapsed during the war, leaving tons of lumber abandoned on the colony's shores. With few ships coming to Gabon, shipping out animals and specimens to the United States was an extremely arduous task.

The French government also imposed a new series of policies that Gabonese people resented. Besides doubling taxes in 1914, colonial officials followed the directive of the Ministry of Colonies to ban the importation of guns and powder for sale to Africans on the grounds that these materials were needed for the war effort at home. French authorities forcibly recruited rural Gabonese men to serve in the military or work as porters carrying material to fight against the Germans. Furthermore, the governor general of French Equatorial Africa instituted a new system of hunting permits in 1915 that required individuals to pay a thousand francs for the right to kill gorillas and elephants; this fee was more than the annual salary of urban laborers, let alone rural Gabonese people during an economic depression caused by the disruption of international trade. Charmarande, a French administrator stationed in Fernan Vaz between 1914 and 1916, committed so many acts of violence that the colonial government chose to remove him from office. To make matters even worse, heavy rains during the dry season of 1916 compounded the hardships of southern Gabonese communities, to the point that a major famine gripped most of Gabon for the next three years.[56]

Garner commenced to explore new opportunities to profit from his Gabonese experience soon after his 1914 homecoming. The famed hunter and gorilla aficionado Carl Akeley joined the New York Zoological Society to form the Garner African Film Company, which aimed to send Garner back to Gabon to film wild animals.[57] When this venture failed to raise the capital needed to send Garner back to Africa, Garner turned to other potential patrons, including the New York Zoological Society and the Smithsonian Institution. On the recommendation of William Hornaday, wealthy Philadelphia businessperson Alfred M. Collins abandoned his plans to hunt tigers in Mongolia and accepted the idea of shooting gorillas in Gabon with Garner.[58] While the NYZS briefly considered and then rejected cofinancing their plans in return for more animals, the Smithsonian decided that the chance to greatly enhance its collections of great apes outweighed doubts about Garner's linguistic theories.[59] A Smithsonian curator noted, "Our exhibition Series of African Anthropoid Apes, probably the most interesting group to the public in any museum, is regrettably and notoriously poor. Without a large sum of money available at all times for the purchase of specimens, it is likely to remain so for a long period of time. The Garner expedition seems to offer a chance to obtain this much needed material."[60] Garner's relationship with both the film company and the NYZS seem to have soured by this time, particularly because of what Garner deemed "the men of great means" who were "so patronizing in their manner that one of my temperament cannot approach them."[61] Smithsonian secretary Charles Wolcott thus secured Garner the backing of the museum.[62] Collins and Garner recruited Professor Charles Furlong to help film their efforts, and the Smithsonian sent Charles Aschemeier, a twenty-three-year-old assistant taxidermist, as their representative.

The expedition began to coalesce into a viable proposition in late 1916, a year after the initial proposal had been made by Garner. As the Smithsonian's finances were in dire straits, it appears that Wolcott and the other supporters of Garner at the museum hoped that Collins's wealth would alleviate their own contribution. The Smithsonian agreed to cover the shipping costs of all specimens from Gabon and to pay Aschemeier's salary and travel costs. Garner had to depend entirely on Collins.[63] The French government agreed to permit their entrance to Gabon, although Aschemeier's Teutonic last name so disturbed the French ambassador to the United States that Charles Wolcott had to prove Aschemeier was not a German sympathizer.[64] Finally, Garner and Aschemeier left New York for France and then Gabon in late December 1916. Collins and Furlong expected to join the others later in 1917. Problems beset the expedition immediately. Aschemeier's ticket and lodging in France proved much more expensive than he or any of his superiors in Washington had expected.[65] Both Collins and Furlong enlisted in the U.S. Army when President Woodrow Wilson declared war on the Central Powers. Once the Smithsonian staff realized how expensive the trip had become, they chose to have Garner and Aschemeier stay in Africa. The unlikely duo spent the next two years in Gabon without regular financial support from the cash-strapped Smithsonian, which eventually became unwilling to furnish additional money for the two men.[66]

The war's impact on Gabonese people and foreign residents of the colony did not escape Aschemeier and Garner's attention. Once Garner reached Port-Gentil in March 1917, he discovered that he had to pay two thousand francs under the new regime of hunting licenses and that prices for supplies like sugar and tea had skyrocketed. Telegraph lines broke down, and boat service from Port-Gentil to Fernan Vaz had become very irregular. French officials at Port-Gentil added insult to injury by not initially recognizing the exceptional status of the expedition. Only the intervention of the governor general of French Equatorial Africa solved this problem, just as he had prior to World War I.[67] The home of the expedition in Ombouè, the administrative capital of the Fernan Vaz district, illustrated the links between the expedition and the patronage of the colonial government. The two men lived for two years in a house normally reserved for French officials.[68]

About the only accomplishment of which Collins and Garner could boast was the number of animals they had managed to acquire. With an undiagnosed kidney disorder that would take his life in 1920, Garner's poor health placed most of the work on Aschemeier to collect specimens.[69] Once the governor general had approved Garner's application for an exception from hunting licenses, Aschemeier began to shoot hundreds of birds and mammals. Garner lauded his skill: "Nimrod (Mr. Aschemeier) is still slaying a few dragons and today accounted for 18 bats. I heard the firing and thought the Kiel fleet was coming into the lake [Fernan Vaz]."[70] By 1919 Aschemeier had sent more than 2,300 animal specimens to the Smithsonian. He kept a meticulous record of the many birds he brought down.[71]

Poor economic conditions and food shortages made the two men popular employers among Gabonese hunters. In Fernan Vaz, famine scoured the land, and even Americans and Europeans struggled to obtain enough food. Prices for locally grown manioc and plantains had gone up over 400 percent between the beginning of the war and 1918, in part because of excessive government regulations and a drought in the fall of 1918.[72] Domestic servants stole food from the Americans and their other African employees or dropped potatoes and meat on the floor, because "no white master would partake of viands picked up from underfoot." Others stole birds and porcupines from Aschemeier's traps to eat: "On such occasions back will come the native with the foot of a porcupine or a *nquani* [bird], and tell almost tearfully how the wild things had twisted off their own feet and escaped. Nor were they ever able to see how neatly the creature had cut off its limb in the twisting process."[73]

Under these circumstances, Fang, Nkomi, and Gisir men needed work, especially with the near-total collapse of exports for ivory and rubber. "A perfect tornado of blacks, many of whom had worked for me before, swept down upon our establishment, desiring employment as 'shoot-men,' house-boys, chiefs, guides, and what-not," Garner wrote, and noted how salaries had risen considerably due to the high cost of food and merchandise. Even Senegalese colonial guards moonlighted as guides for the expedition.[74] A Fang hunter whom Garner had known since the mid-1890s, Donga Njego, traveled about a hundred miles to tell of gorillas located near his home village. Besides hunters and domestic servants, Aschemeier and Garner required help stuffing and preserving dead animals. Their management of these tasks reflected their attitudes towards Africans: "We trained negroes to do this work, but it was necessary for a white man to stand over them with a club to enforce proper attention." Such demeaning treatment did not outweigh the fact that the expedition allowed Gabonese hunters to ignore laws that restricted their access to firearms and banned the hunting of big game without permits. After shooting a gorilla, for example, the Americans distributed the meat to local people.[75] On leaving Gabon, Garner gave out rifles to his favorite hunters, even though this act violated colonial gun laws.[76] Killing gorillas and other game provided protection for fields endangered by wild marauders. Èdembè, the same chief who had worked with Charles Jones in 1914, led Aschemeier to a gorilla so that it would no longer destroy his family's gardens. Chiefs in other settlements also aided the hunts.[77]

Aschemeier became an unwilling participant in indigenous spiritual practices as he hunted, much as Garner had prior to World War I. His observations of a trial of a man suspected of transforming himself into a man-eating leopard illustrate both the willingness of the American researcher to claim to be a representative of the colonial state and the spiritual tensions arising in this difficult period of social and economic instability. Florence Bernault and Christopher Gray have both noted how fears of "leopard men" articulated anxieties among Europeans and Gabonese people regarding the violence of the early colonial period, as well as how leopard-men attacks represented efforts by male Gabonese *mwiri* power associations (secret societies) to obtain power and intimidate others in the face of growing French authority.[78] Outbreaks of purported ritual murders occurred in much of southern Gabon from World War I through the

1940s. Killers, whether shape-shifters or simply dressed in leopard skins, were believed to seek out human sacrifices and body parts to obtain supernatural power.

Aschemeier's encounter with an accused leopard man took place on the island of Anguanamo on the Iguèla Lagoon, where Èdembè lived.[79] Leopards regularly raised their young on this island in the dry season, when the island was connected to the mainland. One day Aschemeier ran across Èdembè with a Fang-speaking man. Soon afterward, a member of the small Ngovè linguistic and ethnic community showed the American a man bound in chains. Ngovè and Fang people accused the prisoner of killing a man while in leopard form several days before. Angry, Aschemeier demanded that the man be released, especially on learning that some Fang-speaking people wished to put the suspect through a painful ordeal involving the use of spicy herbs placed in his eyes to see if he truly had used supernatural power for evil. He ultimately threatened to send colonial guards to the village unless the case was brought to the local administrator's attention. Èdembè and the assembled people agreed to this, most likely out of fear of the violence and theft that often came with visits by guards to Gabonese communities.

Although Aschemeier included this incident as evidence of African cultural and intellectual inferiority, its real value lies in how it delineates the multifaceted strategies of Èdembè. Gabonese priest André Raponda Walker noted this chief had earned a reputation for his mastery of supernatural forces, particularly of powerful *ombwiri* water spirits.[80] More tangible problems also troubled him. The elderly chief had to placate Fang-speakers who did not fully recognize his authority, an American trading partner well-connected to the colonial regime, and Ngovè and Fang people worried about murderous uses of mystical power. Since the turn of the century, Fang clans and Ngovè people had already claimed rights over land and valuable trees.[81]

It did not help matters that the alleged killer was Ngovè, so Èdembè might be suspected by Fang-speakers of sheltering a murderer. When Aschemeier first complained about how the prisoner could easily choke from his chains, Èdembè offered to loosen the bonds but then complained he could not actually find the key to the lock. Aschemeier noted his amusement at hearing Èdembè complain in pidgin English in his search; such comments must have been meant for Aschemeier to hear, because

Èdembè also spoke Ngovè and the coastal lingua franca of Omyènè. Once the key was brought from another chief, the trial began. Èdembè chose to hear the case at some distance from the village where Aschemeier was staying, on the grounds that it might disturb his American guest. A Ngovè man summoned the American to the assembly, where Èdembè asked him for advice about the well-armed Fang delegation's demand for a poison ordeal. Aschemeier then threatened to send for the administrator's guards, and the Fang backed down. Èdembè had used Aschemeier to ensure the safety of the Ngovè man but had not allowed his foreign guest to completely dictate the treatment of the prisoner. Also, he had visibly shown off to his Fang neighbors and Ngovè subjects his own partnership with Aschemeier. Unfortunately, no other vignettes by Aschemeier or Garner offer such a detailed discussion of Gabonese social conflicts during the expedition.

The two men preoccupied themselves with shipping out their specimens rather than investigating how Gabonese communities coped with troubled political and economic conditions. Unreliable transport within Gabon and the lack of adequate places to store specimens hindered their plans. Aschemeier, apparently responding to criticism from Wolcott regarding the poor state of several packages, described the reasons for the delays in February 1919. Three barrels of specimens filled with brine and four other cases had sat in a corrugated-iron roofed building at Port-Gentil for several months, due to the limited number of steamers visiting the Gabonese coast. "You know it is hot here and I imagine they were more or less put through a boiling process," he noted.[82] French military regulations on shipping also impeded shipment. Sometimes, captains chose to first accept and then reject bringing the materials aboard, and then billed the expedition for embarkation and debarkation.[83] When Wolcott complained about the situation to the Chargeurs Reunis French shipping company, the company replied that the government ultimately had final say over what could be brought aboard their ships.[84] Garner had more success with the English Thomas trading firm due to his long acquaintance with the owner, and this company shipped some of the materials as well.[85]

Neither man ever returned to Gabon after their final departure in March 1919. Their return put Garner's name back in the American headlines. Major news and opinion magazines again published his articles on animal life and Gabonese people. However, Garner passed away in

February 1920, and so enjoyed the fruits of his long Gabonese career for less than a year. Aschemeier worked as a taxidermist for the Smithsonian until his retirement in the late 1950s, but his further journeys led him to Brazil rather than Africa. French collectors dominated exotic animal collecting in Gabon for the rest of the colonial period. In Fernan Vaz, professional hunters and animal collectors such as Georges Trial made a living through partnerships with Gabonese communities. Gabonese men continued to circumvent state policies on guns and hunting through their foreign patrons, and French collectors remained dependent on indigenous expertise to obtain animals. Despite the consolidation of the colonial state in Gabon after World War I, the patterns of relationships within exotic animal collecting in the colony as depicted by Richard Garner persisted to the end of the colonial period.

Conclusion

Richard Garner's career collecting animals in Gabon reveals a great deal about the reliance of American zoos on colonial regimes and the everyday negotiations between foreign animal collectors and Africans in the early twentieth century. The onset of concessionary companies, increased tax burdens, and World War I saddled rural Gabonese people with heavy demands on their labor and their resources. As the colonial administration and private companies eroded Gabonese access to hunting and international trade, indigenous traders and chiefs benefited from their alliances with exotic animal collectors such as Garner. The American researcher's Gabonese employees and trading associates obtained weapons, food, trade merchandise, and money during a period marked by economic downturns. Because the French administration allowed Garner to operate freely, he was an attractive alternative to the exploitation of the CCFV firm for Gabonese men, even with his avowed faith in biological white supremacy.

Another important point raised by a review of Garner's interactions with his Gabonese clients is the discrepancy between common narratives of exotic animal hunting in North America and the mundane dealings of animal collectors in African colonies. Garner himself almost never obtained a gorilla or a chimpanzee on any hunt. Instead, he relied on a large network of British and French traders, missionaries, African merchants and political leaders, and the aegis of the colonial government. His reputation as an adventurer gradually drew support from backers at the Bronx Zoo and the Smithsonian Institution, but he hardly fit the model of intrepid heroism that so dominated early twentieth century American representations of hunting in Africa. North American discussions of natural selection and racial differences made chimpanzees and gorillas commodities in high demand, and Garner's connections rather than his hunting acumen permitted him to profit from this business. A sense of camaraderie among European and American men living in the colony, a need for European currency and merchandise among Gabonese hunters, and the eagerness of zoo and museum directors in the United States to purchase rare animals at high prices all added up to profits for Garner. While most studies of animal collectors in colonial Africa have mainly considered how suppliers sought to build scientific reputations in Europe and North Africa, this chapter has demonstrated how animal collectors negotiated with colonial governments and African communities.

Garner's experiences also reveal how American museums and zoos developed alliances with European colonial regimes in Africa. Certainly, there is no question that displays of exotic animals in museums and zoos articulated hierarchies of white supremacy and espoused the regeneration of white Americans through exposure to powerful animals. However, historical research on zoos has tended to ignore animal dealers in colonial Africa, even though further inquiry on these figures would expose the finances and the assumptions of museum and zoo staff. Likewise, historians of colonial Africa would do well to consider the writings and lives of animal collectors documented in museum and zoo archives as important sources on social and cultural change in the early twentieth century. For example, the historiography of Gabon—and indeed on most French territories in Africa, with the exception of Madagascar—has almost never used animal collectors as a window into the everyday lives of Africans in the colonial period.

Finally, this study presents new perspectives on the history of African environments in the colonial era. The French did little at all to promote animal conservation, particularly in comparison to the most consistent and much more well-documented efforts by officials in German East Africa, Kenya, and Southern Rhodesia to regulate hunting and forest use prior to 1914. Aschemeier and Garner shot down thousands of animals with impunity after the establishment of a licensing system, while Gabonese people were expected to abandon hunting many animals because they did not have the means to pay for fees. Although the history of gun regulation in French Equatorial Africa has yet to be written, it can be said that the curtailment of gunpowder imports into Gabon proved to be an inadvertent but effective means of limiting the hunting of big game. Garner, for all of his anti-imperialist ruminations, could thus sell animals thanks to the privileges bestowed on him by the colonial state.

It is hard to say what Gabonese people thought of this obsession with animals in the early twentieth century, but a story told by a priest from Cameroon at the Sainte Anne mission on New Year's Eve 2009 suggests how sales of animals could articulate fears about the demand for natural resources in the region. He decided to explain why there were such hard times in central Africa through a story. Once, a white man came to a village and wanted to buy some monkeys, and he paid 5,000 CFA (roughly 10 USD) per animal. The whole village brought in monkeys, and the white man doubled his price. Soon, so many monkeys had been captured that their cost doubled again. The white man announced that one of his brothers would set up a store in the village and then left. His brother came and announced a great deal. He would sell the monkeys back to the villagers for 150 USD each, and the Africans could be assured that the first white man would return and buy them back for 300 USD. Every family bought the white man's monkeys. The store then closed, and neither stranger returned to town. Everyone had purchased the monkeys they had originally sold. That is why there's no money here, the priest concluded. Natural resources like animals might be valuable to foreigners, but the ultimate costs lie with Africans unable to set the terms of trade. This loss of control began in Garner's day, and it has yet to end.

CHAPTER THREE

Is the Monkey Man
Manly Enough?

FEW SONS OF SOUTHERN APPALACHIA in the nineteenth century traveled as far from their original homes as Richard Lynch Garner. Following a conventional path in his early years during the tumult of the Civil War and Reconstruction, he served in a Confederate regiment, spent time in a prison camp, and then worked as a teacher and a real estate speculator until he commenced his African career. Like other southern middle-class men of his generation, Garner struggled with the legacy of defeat and the reality of economic underdevelopment compared to other parts of the country. The rhetoric of economic and educational development in the New South strongly appealed to Garner, who considered himself to be a loyal southerner and a member of an international scientific community, despite his lack of a university degree. However, Garner's disdain for feminists and African Americans hardly made him stand out from the crowd.

Over the course of the 1880s, Garner slowly steered away from this predictable path and embarked on a series of transatlantic journeys. Garner's hope of becoming a gentleman-scholar ran aground on the growing professionalization of scientific research in the late nineteenth century. To overcome the lack of enthusiasm that academically trained scientists at the Smithsonian Institution showed toward his theories, Garner followed the path of Paul Du Chaillu, the famous Franco-American explorer who had claimed to be the first Westerner to hunt gorillas in Gabon. Again and again, his critics portrayed Garner as a fraud who demonstrated his lack of proper manliness by lying about his improbable exploits. Garner struck back by portraying himself as a brave man willing to sacrifice his own name for scientific progress. By contrast, he contended that his opponents lacked the fortitude to conduct scientific research in Gabon and remained so bound to conformity that they could not appreciate the brilliance of Garner's research. White manhood thus became a means of determining credibility in transatlantic commerce and intellectual questions.

Garner harnessed popular narratives of self-sacrificing scientists and the regenerative possibilities of colonial adventure to overcome his marginal background as a southern Appalachian middle-class intellectual. Atheism, scientific progress, and African expertise all allowed Garner to cut himself free of his earlier background to become a nationally respected figure. His life sheds light on eccentric examples of southern manhood, a rare topic given that most studies of masculinity in the Gilded Age and the Progressive Era focus on the northern elite and middle-class men.[1] While the literature on masculinity in the antebellum South is extensive, relatively few studies have considered the changes in manhood in the late nineteenth century.[2] Glenda Gilmore, one of the few historians to seriously consider masculinity among white southerners in the New South, has made an observation that applies to Garner: "If victorious and sober Yankee men questioned their own masculinity in a rapidly industrializing and urbanizing culture, southern white men added a loser's shame and degrading poverty to that burden. Despite the New White Man's eagerness to be on top—to rise economically by exercising self-mastery—his talk was democratic. He might have been a bit of a rube, but he was a smart one. He might have started in a humble home and traveled far, but he never forgot his origins. His road, however, led only one way: to town."[3]

Garner might have made an idiosyncratic career for himself, but his anxieties as an Appalachian town intellectual were hardly unique. Garner idolized individual achievement, and thus he joined many of his southern peers in adopting the northern ideal of self-made men for themselves rather than remain shackled to the Lost Cause.[4] Just as Appalachian townspeople from the Civil War onward increasingly demeaned poorer and rural people as rednecks and white trash, so did Garner wish to place himself in the forefront of scientific innovation and the national mainstream rather than be seen as an exotic curiosity in need of outside help.[5] Before turning to Africa, Garner himself had written "local color" stories that highlighted Appalachian dialect and folk culture as radically different from typical (and northern) American society. By acting as a cultural ambassador for his region in ways that distanced himself from the backward connotations of Appalachia, he prepared the way for his later efforts to show off his intellectual capabilities and scientific training in contrast to supposedly superstitious and irrational Africans.

Garner's published work demonstrated pride in his individual bravery and scientific successes, but his diary and his letters to his son display his profound insecurities about his worthiness as a man. Garner had abandoned his wife for Gabon. His rootless life moving from central Africa to various northern cities did not allow him to act as a caring supporter of his family. Garner continually advised his son to make money and to buy a home, instead of chasing foolish dreams of intellectual acclaim. His misguided effort to woo Ida Vera Simonton, a New York socialite who visited Gabon in 1907, again reminded him of his failure to maintain a lasting relationship with a woman. Garner's efforts to present himself as a caring and protective mentor completely alienated the free-willed Simonton, who rejected conventional ideas about marriage and male dominance. His unwillingness to engage in sexual activity with Gabonese women further separated him from both African and European men living in the colony. Garner's intimate disappointments underscore how public performances of virile manliness could mask profound anguish. Ironically, Garner eventually found patrons at the Smithsonian Institution and the Bronx Zoo after 1910. His new allies praised Garner as a tough-minded and self-made genius, precisely the figure he wished to be, instead of the melancholy failure he so often believed he had become.

Garner's case offers several important contributions to a historical understanding of masculinity in the Gilded Age and Progressive Era, as well as the connections between constructions of manhood and American portrayals of Africa and empire. His ideal of manly achievement was not a matter of Old South nostalgia or a celebration of Anglo-Saxon Christianity, even though he passed through the crucible of the Civil War and the dislocation of Reconstruction. He presented himself far more often as a global citizen of the scientific community than as a southerner. Scientific research, a celebration of technological change, and African travel allowed Garner to transcend his previous background and (theoretically) his inability to maintain his marriage or act as a supportive father. The literature on masculinity in the Progressive Era has contended that many men tried different ways to tap the supposed vitality of tropical peoples and environments to renew manhood damaged by impersonal bureaucracies and industrialization. Garner demonstrated his self-mastery as well, but his case indicates how far less prominent individuals could appropriate similar discourses of empire building and manly struggle for their own ends. Furthermore, the chasm between private fears and Garner's detractors also presented gendered arguments, as they questioned Garner's veracity and his claims of intrepid self-sacrifice more than they objected to his actual theories. Manhood became an aspect of social and cultural capital that could literally make or break a scientist's reputation. His final ascent into respectability came only after well-known northern scholars decided Garner was a real man after all, even if they did not endorse his ideas regarding animal communication.

The Making of an Eccentric Southern Man: Gender and Richard Lynch Garner's Early Career, 1848–1892

Garner's early life prior to the Civil War appears to have been fairly happy, at least as it is described in a short biography Garner wrote from 1900 to 1904. His early models of proper manhood reflect his somewhat distant relationship with his father, Abingdon's role as a local commercial center, and the ubiquity of slaves. Garner briefly praised his father as an enterprising businessperson, but he appears to have not had much daily interaction with him. Garner's father mainly appears as a disciplinarian who had to tame his son's antics. "My father gave me one of those 'spare the rod and spoil the child' exercises, which, in those pious times, was supposed to satisfy the divinity who presided over truancy and mendacity." Garner's older brother was not much more inspiring—he scared his younger brother with ghost stories, beat him up on occasion, and loved embarrassing the young boy with pranks at his expense.[6]

These interactions posed Garner with multiple ideas about race. One slave preacher's public-speaking ability and kindhearted demeanor so impressed the boy that he briefly considered becoming a pastor, which was a source of condescending amusement to Garner in middle age: "Uncle Charles, that dear, old friend of my childhood, will ever remain an idol in the temple of my memory" (11). When he had an illness, Garner turned to a central African–born slave healer to provide him with an herbal remedy

(56–57). Yet such interactions did not lead Garner to question his family's own ownership of slaves. Garner himself recalled becoming enamored of "legerdemain and negro minstrelsy" (40). A girl vying for Garner's affections told him to forget about one of her rivals, because "Becky lived in a very small house and they had no niggers and horses and things" (20). At a circus, an acrobat amazed Garner in ways that reflected his previous attendance at the execution of several slaves: "One man climbed to a bar, swinging high in the air, and there he hung himself by the toes and the heels and finally by the neck and it didn't kill him. This surprised me for I had seen two negroes hung and it killed both of them" (17). Like many other white southerners, Garner held kind memories of individual elderly, dutiful African Americans from childhood, while still holding people of African descent in disdain.

Performers, preachers, and other interlopers passed through Abingdon in the 1850s. Garner found these outsiders to be a source of entertainment and new possibilities, as did many other southern spectators.[7] Garner was so taken with the acrobatic feats of members of several visiting circus troupes that he longed to follow their example (15–16, 29–30). "No city boy has an adequate conception of the thrill of a country boy's heart in carrying water for the animals or helping to raise the tent of a country circus," Garner declared (30). Such visitations offered a break from routines that Garner hated. He did not show much interest in the rigid conformity of his schooling. Religious education filled him with loathing for its demand of lockstep obedience and rote memorization. Sunday for Garner "was a day of inquisition and I dreaded to see it approach" (68). The lures of copious meals and the respect of crowds lured Garner far more than the Methodist teachings espoused by his parents. When a pastor came to dinner, "every member of the household was put on his or her good behavior, and children had to wait and eat at a second table. You can imagine the chance of a boy getting any part of the chicken except the neck, or a wing or occasionally a hindleg that the preacher failed to notice" (72). Instead of mocking his parents as ignorant, Garner contended that they suffered from the same lack of curiosity that led so many people to follow Christian beliefs without much reflection.

Women also entered into Garner's search for admiration. According to his autobiographical account, Garner became rapidly infatuated with a series of young women. His efforts to impress them through his acrobatic antics and his charm did not succeed particularly often. On the other hand, his mother was the center of affection in the family. If one can trust Garner's published account, she followed contemporary morays of the Christian homemaker who lavished affection on Garner. Because Garner wrote only sporadically about his marriage and his intimate relationships with women later in life, it is hard to determine how much his romantic career was influenced by his upbringing.

Compared to his voluminous writings on scientific and African topics, Garner made little mention of his defeats on the battlefield and his family life. He left only terse references to his wife, Maggie, or any aspect of his life from the Civil War through the 1880s. His wife outlived him, but their relationship seems to have almost entirely been severed once Garner commenced his public career. "In my marriage, I may here confess that my highest ambition was to rear a family after an ideal fashion and nothing else on earth can gratify my wish. It is now far, far too late for me to continue hoping . . . and the failure for me to do so has been the sharpest persecution of my life—It has been the cause of my being far less fatherly to the only child I have. . . . For certain reason [*sic*] which might have been and could have been avoided at the most productive time of my life I became a wanderer and have lived a gypsy life for more than 25 years," he wrote on the thirty-third anniversary of his marriage in 1905 while living yet again in southern Gabon.[8] The sense of personal loss that pervades this passage also comes across in his lengthy correspondence to his son, Harry, and his daughter-in-law. Time and again, Garner urged his son to settle down and buy a home—precisely what he himself did not do. For Garner, home ownership and a mutually supportive couple formed the basis for any successful marriage.

What went wrong? Why did Garner turn away from such a common late nineteenth-century model that so prized the private sphere of the loving home over a public domain rife with conflict? How Garner's dissatisfaction with his family life and teaching career first emerged is unclear, but he began to pursue a range of eclectic intellectual interests based on his commitments to scientific research and atheism. At some point after he finished his studies in religion and medicine at Jefferson Academy, Garner had rejected the faith of his family: "To my mind the Christian religion is one of the most shocking stories ever recounted. How it survives the light of day is a problem of physics that I cannot solve and I frankly admit that

I regard the teaching of it as a crime of civilization. It is a propaganda of falsehood, a nursery of hypocrisy and night-mare [*sic*] of human progress," Garner wrote (72). Although scholars investigating the rise of American atheism in the late nineteenth century have mainly treated free thought as a northern movement, a minority of white male southerners also became disenchanted with religion.[9] Robert Ingersoll, the most famous American atheist of the late nineteenth century, provided Garner with a master narrative of heroic believers in human progress pitted against deceitful clergy and deluded believers.[10] Ingersoll's celebration of individualism and science as a means of overcoming mental barriers to success corresponds extremely well to Garner's aspirations to become a well-respected scholar regardless of his lack of formal education.

While Garner stayed relatively clear of the political turmoil that led to the triumph of Jim Crow legislation in the South, he did follow the example of many men in the New South by highlighting individual achievements and economic development. For Garner, achievement could be best demonstrated by personal integrity and wealth, rather than by family ties, religious piety, or intellectual pretensions. The booming demand for land in Roanoke furnished one venue for personal advancement. His letter to a professor at the Smithsonian in 1900 asking for funding included a short summary of his financial acumen. After describing a long legal dispute between himself and a rival company president in 1896 over land, Garner recounted to Holmes that his former rival asked him to settle an argument two years later: "Mr. Garner, we have had some very unpleasant relations in business. . . . I will state plainly to you that you are one of the most stubborn men in a quarrel that I have ever had to contend with . . . [but] I never knew you to be guilty of a dishonest or unmanly act. . . . I would be willing to trust you with all the money in Dayton." Garner added that he could fill pages of testimonies by his employers about his honesty and skill making money.[11] During the nadir of his professional fortunes in 1909, shortly before he finally received the patronage of the New York Zoological Society and the Smithsonian, Garner advised his son to concern himself with becoming wealthy rather than pursuing knowledge for knowledge's sake. "Get money and keep it—The theory of working in the interest of science or humanity is not put into practice except by fools—I am surprised that I have been so blind and stupid all my life as not to see every man, preacher, missionary and philanthyopist [*sic*] as well as the alleged self-sacrificing scientist demands money for every effort he makes."[12] Money and its management proved to be crucial features in the ethos of rugged individualism in Garner's life, as they did for many northern men in the late nineteenth century.[13]

Garner's plans to harness the phonograph did not impress potential patrons at the Smithsonian and other institutions. Gregory Radick has noted how staff members of the Smithsonian Institution found Garner's plans vague and unimpressive. In 1892 Frank Baker, a staff member at the Smithsonian, denied Garner had any professional credentials: "He was first a peddler of trees and nursery stock afterwards a real-estate drummer [sic]. Do not understand that I consider these occupations in any way detract from his merits as a citizen or diminish his general worth, I merely think that they are not exactly those that enable one to 'catch the eel of science by the tail.'"[14] Baker assured another professor that Garner was a dubious character who had been "self-deceived" into believing his experiments had any value and that he wandered "from the facts." All in all, Garner's announcement that he would record monkeys and apes in an iron cage in 1892 struck Baker as something that "sounds [more] like Jules Verne than it does proper scientific experiment."[15]

Such critiques aimed to dismantle both Garner's pretensions as a researcher and his claims to be a trustworthy observer. To receive the patronage of wealthy individuals and institutions, Garner had to convince potential backers that he had the proper manly characteristics of good judgment and foresight, along with the ability to stop wild ideas from skewing his research. Garner wooed potential patrons with his espousal to self-sacrificing research, which weaved together a vision of intrepid manhood willing to embrace danger with service to scientific progress. Rebecca Herzig has noted how self-sacrifice was coded in ways that made scientific research the exclusive domain of properly masculine white men.[16] Just as important, Garner's findings struck a chord with the press eager for any information on a "missing link" between humankind and monkeys in an age where Social Darwinism had so excited popular attention.[17]

Garner certainly was no stranger to self-promotion, and he sought to present himself as a gentleman-scholar who fulfilled the proper requirements of manliness. He and his agent, Samuel McClure, sought out introductions with many well-known writers and business leaders, and Garner could proudly show off his new fame by listing his patrons who "have opened their purses, as they have opened their hearts." Among others, his patrons included inventor Alexander Melville Bell and Colonel John Hay (the future secretary of state); author William Dean Howells and Thomas Edison also publicly endorsed Garner's venture.[18] The Virginian celebrity lectured to men's clubs in New York to show his own personal qualifications as well as to expound on his theories.[19] And Garner's 1892 book, *The Speech of Monkeys*, gained a wide audience.

Garner's public persona was best captured in his 1892 summation of his African plans published in the leading journal the *Independent*. This discussion had a detailed set of descriptions of the numerous electric machines at his disposal: telephones, waterproof cable wire, lights, batteries, and a phonograph that would be guarded in a large iron cage in the midst of "primeval Africa." On top of this arsenal, Garner also intended to arm himself with a rifle and a curious device that would silently kill any serious threat by firing an arrow containing toxic prussic acid. He assured potential skeptics that he indeed would put his life on the line for progress: "So strong is my conviction of my ability to learn their language, that I undertake this journey at my own expense, with my life in one hand, and my equipment in the other. Impelled by an honest zeal . . . I shall seek the depths of the unbroken forests, and wrest the secret from those man-like creatures at the threshold of their own abode." Such valiance would not be marred by money, as Garner declared he would rather die in poverty and succeed than fail and gain "the wealth of India."[20] Such high drama allowed Garner to seize a place at the forefront of evolutionary theories and state-of-the-art technology. However, the results of his journeys to Gabon from 1892 to 1908 severely damaged his reputation as an indefatigable researcher. Whether or not Garner met the qualifications of self-sacrifice and honesty became a topic for debate. Garner's defense of his own manhood would rest on his ability to command Africans, even though he depended far more on his Gabonese associates than he admitted to his English and North American critics.

King of the Gun, Fool of the Jungle?
Garner and His Critics, 1892–1909

Garner's first visit to Gabon impressed relatively few Europeans or Americans. As Gregory Radick has noted, the first critics of Garner's portrayal of his Gabonese stay drew their information from several English traders, French administrators, and Catholic priests living in Gabon. Albert Veistroffer, a French officer who governed Fernan Vaz between 1897 and 1900, recalled Garner as an entertaining "American or English" buffoon who did not stray from the grounds of the Sainte Anne mission, contrary to his declaration of being all alone in his iron cage. "No one took him seriously," Veistroffer added.[21] A French missionary remarked how Garner had made himself a comfortable home at the mission and dared stay in the surrounding rainforest only for several days. He reported that Garner claimed he came to Africa mainly to make money, did not actually care if monkeys could speak, and said no one would ever bother to come to Gabon to verify his observations.[22] When African American traveler Charles Spencer Smith came to Libreville in 1894, he noted, "There are a great many amusing stories told regarding Professor Garner's movements. It is said that he pawned his steel cage in Libreville, and he did no more than merely make a show of effort to carry out his original designs."[23]

Garner later presented himself as an honorable exception to the small and venal crowd of European traders and officials who lived in the colony. How different he really was is hard to say. Case in point—Garner assured his son he never had sexual or romantic relationships with Gabonese women. Such liaisons utterly disgusted Garner as a gross violation of the color bar.[24] In theory, this made Garner a member of a very small minority. Practically all English and German traders as well as French officials entered into sexual and romantic relationships with Gabonese women.[25] Colonial officials and travel writers celebrated Gabon as a colony of sexual opportunity, just as Catholic and Protestant missionaries bewailed extramarital sexual relationships as prostitution. Administrator Veistroffer, for example, had a child with a Nkomi woman, even though his only veiled reference to his son in his memoirs was a photograph with the caption, "a Gabonese woman, who had one child with a European."[26] Yet a collection of Garner's photographs donated to the Smithsonian indicate he participated in a favorite pastime of European men: the consumption of erotic images of African women. He owned half a dozen photographs, mostly staged in a studio, of seminude female women.[27] They include two individuals in attire typical of Fang wives, including heavy brass bracelets. Whether or not these images were merely meant to be illustrations for ethnographic lectures or were also objects of private interest is a matter of conjecture.

Garner's early determination to separate himself from his peers waned over time. Over the course of his repeated stays in Gabon from 1893 to 1919, he developed close ties with English and German merchants and French officials and traders. Like almost every European or American in Gabon, Garner greatly enjoyed hunting and celebrated his command over African fauna and people. While Garner privately considered French rule to be unjust and extremely oppressive, he relied on French administrators to provide him with animals, gun permits, and protection. Garner also regularly dined and called on passing traders, administrators, and even missionaries. In the very small social world of white society in Gabon—which numbered only several hundred Europeans before World War I—Garner ultimately found acceptance. While Garner's academic critics lambasted him in England and the United States, Europeans living in the French colony seem to have been more amused than dismayed by his claims to scientific expertise. More important, Garner benefited from the same privileges as other settlers and was even granted the right to trade independently in southern Gabon despite the concessionary company system.[28] Garner may have been an odd individual, but he still ultimately endorsed white supremacy and colonial authority publicly, despite his misgivings about the harsh policies enacted by the French government. Garner's endorsement of common transatlantic tropes of white masculine heroism ultimately allowed him to fit into U.S. and Gabonese hierarchies of power.

Academics and journalists in Europe and America proved far less accommodating to Garner upon his return than European residents of Gabon. The English journal *Truth* published a series of damning articles in which Garner was presented as a first-class fake who hardly moved from the confines of the Catholic mission and left behind his cage and an unpaid bill of twenty pounds sterling with the priests.[29] While McClure's advertising campaign spoke of "Professor Garner," his detractors mocked the title. The *Fort Wayne Journal Gazette* dismissed Garner in typical fashion in 1896: "And now comes the London *Truth* with a whole broadside and a half leveled at our professor of monkey talk. . . . Garner can sauce back in monkey lingo and *Truth* will have to take it. All the same we have no faith in Professor Garner."[30] The New York newspaper *Town Topics* reported that Garner supposedly had threatened to hit the editor of *Truth*, Henry Labouchere, yet had not actually proven his courage. It added, "How could 'Professor' Garner lecture before scientific societies in London and be received with respect while such printed accusations confronted him and his chivalrous threat remained unfulfilled?"[31] Although Garner found backing from some Chicago patrons for another trip to Gabon in 1895 and 1896, the allegations of fraud had so undermined his heroic persona that he eventually had to depend on his own limited resources instead of wealthy patrons. In 1900, he grumbled to a Smithsonian Institution employee, "The vast majority of men who follow science as an occupation are mere mechanics and they are the ones I offend, but they do take umbrage at my way of renouncing the orthodox ways. Heresy is the only true religion has been my creed [*sic*] these many years and I have lived up to it."[32] Garner thus asserted his individual pride in the face of blind conformity.

The Virginian's long struggle to return to manly respectability by popular standards in North America lasted over a decade, and it did not come easily. From 1904 to 1909, he stayed in Fernan Vaz without any sure source of external funding. References to Garner's adventures in U.S. newspapers significantly declined by 1904, and his desperation led him to develop a wild new plan to demonstrate his heroism to the American public two years later. Sometime around 1904 Garner had met Ida Vera Simonton (1871–1931), a wealthy New York socialite who dabbled in journalism. She made plans with Garner in late 1905 to visit the following year but did not inform the press until her departure for Africa in the summer of 1906. Ostensibly, Garner asked her to come help edit three books but told her not to disclose any details of her visit to the press.[33] Simonton agreed, largely to promote her fledgling career as a writer. Garner spent most of his remaining resources on the renovation of his house for Simonton. However, he expected to be richly rewarded for his efforts.[34]

He had good reason not to inform Simonton of his real plans. The French government's creation of a poll tax in 1902 and the economic recession that came with the concessionary companies' rapacious policies had led some Gisir and Mitsogo clans in southern Gabon to rebel against the French government. Garner decided to use Simonton to end the revolt, abolish the concessionary system, and become an international sensation. He boasted to his son, "I expected this proceeding to give her such notoriety and prestige as to create a demand for her on the lecture stage and as for her writings on Africa in all of which I expected to share and while I confess to mercenary motives as a part of the scheme, it was wholly legitimate and feasible." In a weighty forty-page letter to his son, Garner described his plan. "Having always entertained the kindest of sentiments towards [Simonton] and having confidence in her abilities as an author and critic, I was glad of an opportunity to form an alliance. . . . I was conscious that it lay in my power to do vastly more for her than I had ever promised," he declared.[35] He decided to make her a queen by convincing Gisir or Mitsogo leaders to appoint Simonton as *oga nwato ntangani*, or "female white chief," in coastal Gabonese lingua franca of Omyènè. She then would convince the French government to abolish any special privileges to the hated concessionary company monopolies in return for an end to the revolt. Garner did not speak a word of the rebels' language, and Mitsogo politics revolved around village and clan councils, but yet he insisted he somehow could pull off Simonton's coronation. Such a fanciful plot could never have worked even if Simonton actually agreed with it, but Garner quickly found she had her own ideas of adventure.

Simonton's own search for personal emancipation violated Garner's thorough contempt for feminism. A long screed he wrote on the issue of women's rights asserted that coeducation and suffrage would only coarsen and debase women's purity. "The man who admires masculinity in a woman is deficient in that quality in himself and masculinity is the only result that can follow a masculine life. In my opinion, no real masculine man can tolerate masculinity in a woman," he warned.[36] Feminists were merely "ugly old maids" who could not find husbands, and they preferred to accept "women's privileges" instead of complete equality. Nature had ensured that males of every species in existence always dominated. Simonton, though, would end up the winner in her contest of wills with her cranky host.

Garner presented his hapless relationship with Simonton as proof of his chivalrous manners. She preferred to drink and travel with English traders and French administrators instead of following around monkeys or ensuring her enthronement as queen of the Mitsogo. English, French, and German traders in Gabon had long celebrated violence toward Africans, heavy drinking, and sexual relationships with coastal Gabonese women. Although Garner's antipathy toward Christianity hardly made him a friend to pastors and priests, he shared their conservative views regarding sexuality, particularly when it involved crossing racial boundaries. English traders were not good gentlemen: they were "more interested in the pleasure of her company than protecting her name from scandal."[37] Garner's fear of physical and moral degeneration of Europeans in colonial Africa was hardly unique, and neither was his self-portrayal of a seasoned paternal expert seeking to guide a wayward single woman.[38] He assured his son that he had sworn to Simonton's brother that he would guard her honor with his life, but Simonton had no intention of being a damsel in distress in need of Garner's manly protection. Yet he learned from his Gabonese canoe workers that a trader had claimed Garner had brought Simonton out for "improper purposes and that now that she was here the trader in question had induced her to drop [Garner] and become his concubine." Whether or not Simonton actually had become the trader's lover, she chose to stay away from Garner as much as she could and "laid down the law to me as if I were her menial instead of her benefactor" while they were together.[39]

Simonton spurned Garner's effort to remake himself into a manly romantic hero, leading to one uncomfortable incident after another. On a visit to the Atlantic port of Cape Lopez, Simonton and Garner ate at a trading house. After everyone retired for bed, Garner heard a man's voice in Simonton's room, rushed over, and confronted the bookkeeper of the post chatting with his female guest. A brief fight ensued, apparently to the annoyance of Simonton, who blamed Garner. Garner linked race and his southern identity to his persona: "Now, my son, I need not remind you of the sentiments of your own race and country when a woman under your protection is approached however adroitly by any man. Miss S. was reared in a different school."[40] At times, Garner contended she was merely naive about the knavish traders, but he eventually decided she was culpable as well. While she turned down Garner's offer to kiss her, she became infatuated with a young Italian trader who "she suspected of being a young nobleman incognito — as any 16-year-old girl would be about an actor or a soldier."[41] Perhaps his vexation was especially galling because she slept in his bed and occupied his house while he was stuck sleeping on his veranda.[42] Her refusal to kiss him one day in Gabon might have also had something to do with Garner's anger (39). Simonton asked Garner's household servants about prostitution and sexual relations between Europeans and Gabonese women, which also appalled her host (29). By

June 1907 they no longer had any plans to write together, and Simonton blamed the lackluster results of her stay on Garner's incompetence (32). He told his son how he could have landed her a hundred honors and that he was the "only white man in the colony the governor himself not excepted" who could have pulled it off (35).

Garner portrayed Simonton as a deviant: "She is one of the most unwomanly women that I have ever known—She has an idea that rudeness and bravado, insolence and vehemence [*sic*], coarseness and invectives are equivalent to courage and talent, but as I know her she is simply a masculine female."[43] As their partnership collapsed, she told Garner that editors and everyone else who knew of Garner "had said the most cruelly harsh things [about him] the human mind could devise" (30). Simonton had accused Garner of making sexual advances upon her, which he strenuously denied (32–33). The woman "boasts at 30 years of age that she had never loved a human being and yet has the vanity to believe that she has been the idol of a hundred men of the most exalted ranks," Garner groaned in disgust (37). Further charges followed throughout 1907 and 1908, well after she left for Leopold II's Congo Free State and the United States. Simonton's host derided her poor housekeeping, revealed that honesty is "as foreign to her nature as hair is to soup," and reported that she stole photographic negatives.[44]

Garner's plans had flopped yet again, leaving him plenty of time to ruminate about his personal failures. An existential crisis emerged, as Garner sought to reconcile his individualism with his need for help. On the one hand, selfless devotion to scientific progress allowed him to imagine himself as a man superior to others. After he began one essay by noting, "Today the last particle of my manioc is finished [and] I have not one cent to buy anything with," he then fumed about "certain men have title who have gained their notoriety solely through the selfish, sordid, and basely plebian occupations of making money."[45] A mysterious group of Philadelphian financial backers and missionaries both worked to thwart his plans, thus allowing Garner to take on his typical role as a proud maverick. "I will here state that before I shall ever allow [the Philadelphia syndicate] to profit one penny from my suffering, I will starve and die right on this spot," warned one piece.[46]

By 1909 he slowly managed to salvage his career from the Simonton fiasco. To stay solvent, Garner wrote articles about his African adventures.

Most were never published, but some did make their way into the pages of *Sports Afield* and various newspapers. In these articles, Garner posed as the embodiment of trenchant intelligence and daring action. Africans alternately served as a backdrop to his individual achievements, whether through their adulation of Garner's feats or as ignorant foes that Garner could overcome. An Orungu village had celebrated Garner's killing of a dangerous buffalo by calling him *oga njali*, the gun king.[47] In another story, Garner recalled how he evaded being held hostage by Fang-speaking warriors who had tricked the researcher into a trap with talk of a captured live gorilla.[48] Garner's declarations of valiant manhood eventually began to win over some notable patrons.

The Manly Resurrection of Garner's Career

Despite his critics and his own missteps, Garner saw his reputation reach new heights in the last decade of his life. Although he failed to convince academically trained scientists and the general public of the value of his theories of animal communication, Garner convinced leading museums and zoos to bankroll four separate expeditions from 1911 to 1919. His articles again graced major news magazines such as the *Century* and the *Forum*. Garner's self-representation as a brave maverick able to master African people and animals convinced leading natural science institutions that he was a gentleman-scholar worthy of patronage. His ability to successfully bring rare chimpanzees and gorillas intact to the Bronx Zoo and the Smithsonian Institution further attested to his heroism, at least for his backers.

Garner's campaign to prove his manliness commenced upon his return to the United States. He launched a highly publicized tour in 1910 with Susie, a chimpanzee with whom he claimed to be able to hold a conversation. On arrival, a *New York Tribune* article heralded Garner as a "modern Ulysses" and a "modern student of human and animal nature, with the patience of the proverbial Job."[49] The media sensation following the tour won over an influential promoter of the natural sciences: Bronx Zoo curator William Hornaday, an old critic of Garner. Even when Garner introduced his Susie to other apes at the Bronx Zoo in 1910, the two men had not yet met each other.[50] Yet their collaboration would help Garner finally attain the national respect as a naturalist that he had craved for so long. In the process, Garner relied on his new patrons at the Bronx Zoo, Gabonese hunters, and colonial officials.

Thanks to his voyage, Garner had made his way into the good graces of the wealthy elite who belonged to the New York Zoological Society, even though some doubters remained. Hornaday told Garner, "A gentleman prominent in the Zoological Society has wagered me a dinner that you will not secure a thousand dollars worth of animals. Of course I immediately took up the wager, and I expect to win."[51] He then commissioned Garner to return to Gabon to buy gorillas in 1913. After a year and a half, Garner chose to come back to the United States with his gorilla, several chimpanzees, and other specimens. The onset of World War I made his return a difficult one, but Garner managed to impress the Bronx Zoo's directors yet again with his determination. The NYZS granted Garner a lifetime membership in their organization on account of his services.[52] Hornaday wrote after Garner's death, "Of all the men whom I have known or read, the late R. L. Garner knew by far the most of gorilla habits and character by personal observation in the gorilla jungles of equatorial Africa. And never, in several years of intimate contact with Mr. Garner did he so much as once put forth a statement or an estimate that seemed to me exaggerated."[53]

Moving up the ladder of respectability led Garner to move farther and farther north, away from his southern background. Garner had moved to Philadelphia by 1903, and the 1910 U.S. census gives a boarding house as his place of residence. Between 1911 and 1914 Garner again lived in Philadelphia, before he relocated to New York City by 1915. Northern cities furnished Garner with better access to public and private financial backing than eastern Tennessee or Kentucky. Garner's thoughts on urban living were few and far between, although he did tell one reporter, "The very people who say my work is dangerous dart every day in front of speeding automobiles and jam themselves by the hundreds of thousands into subway cars. Say, I'll back the jungle against New York subway cars for safety any day."[54]

Garner's stays in Philadelphia, New York, and Washington paid off as he struggled to prove his character. Ales Hrdlicka (1869–1943), the director of anthropology at the Smithsonian, became impressed with Garner's mettle. Hrdlicka remembered Garner as "an indefatigable worker, and [one who] has contributed greatly to our knowledge of the life and habits of the apes."[55] So did John Harrington (1881–1961), another Smithsonian anthropologist who had first become interested in ethnographic research from reading Garner's books in high school; decades later he remained deeply impressed with Garner's "experiences afar in the jungles in Africa."[56] Alfred Collins, the wealthy president of a paper manufacturing firm and director of the Geographical Society of Philadelphia, was persuaded by Hornaday to form an expedition to Gabon in 1916 under the auspices of the Smithsonian Institution.[57] Charles Furlong, a big game enthusiast from a well-to-do Boston family, joined the expedition as well. Garner basked in his ascent and boasted to Furlong, "I have the vanity to believe, and the cheek to say that I know more people personally and by name . . . than any other one man living. . . . When I first commenced my scientific work, it was entirely as a fad of my own. . . . At this date I am flattered by the fact that every great University on both sides of the earth has taken a light from my torch and instituted a department of animal psychics."[58] While none of Garner's colleagues and benefactors commented much on his theories, the Virginian's heroism and self-sacrifice had allowed him to attain the respect of other scientists. Although Collins and Furlong eventually served in the U.S. military instead of joining the expedition, Garner stayed in Gabon from early 1917 through March 1919.

It would be a typical irony for Garner that he died alone and unrecognized in a Chattanooga hotel. After his final return from Gabon and in between lecture engagements, Garner had slowly begun to collect his writings into several book manuscripts. He had developed a new scheme to study primates at a Havana zoo and perhaps establish a research facility in Florida. After lecturing to the National Geographic Society in December 1919, he spent the Christmas holiday with relatives in Bristol, Tennessee, before setting off for Florida by train. After stopping in Chattanooga, he checked into a hotel, where he passed away. Garner might have abandoned the South, but an Abingdon cemetery became his final resenting place.

Garner's reputation declined almost immediately after his burial. While a series of Garner's articles reached the public in the *Century* and the *Forum*, no completed book manuscripts accompanied them. Robert Yerkes, one of the leading primatologists of the first half of the twentieth century, summarily dismissed Garner as a poor and untrained observer whose theories held no value: "His publications indicate serious lack of scientific competence."[59] Scientists who stressed formal training and a willingness to firmly ignore African local knowledge had little use for the Virginian. His tales of heroic adventure and self-sacrifice for science might have persuaded wealthy patrons to support him in the early twentieth century, but as specific details about his career faded from the popular press, only his unpopular theories remained in view. Researchers who conducted their work in controlled settings with institutional backing could point to Garner as an entertaining amateur. Garner's disinterest in highlighting his southern Appalachian background also helped ensure his future obscurity. Neither academics at southern universities nor partisans of southern or Appalachian identity paid any attention to his career. Garner's last years of redemption through valiant manhood still left his reputation exposed to academic criticism, and his choice to downplay his origins left him isolated from potential supporters.

Conclusion

Richard Lynch Garner's life reflects a series of challenges faced by southern and Appalachian intellectuals, as well as by researchers who lacked high levels of formal education in the Gilded Age and Progressive Era. Like so many white southern Appalachian men of his generation, Garner experienced the dislocations of the Civil War and Reconstruction. His ideas regarding race, family, and adventure were formed in the 1850s, just as railroads transformed southern Appalachia into a major link in regional and national trade. Garner's aspirations of social advancement led him to distance himself from his origins and their reputation for backwardness. Other southern townspeople employed New South rhetoric to justify industrialization and the endorsement of rugged individualism; Garner chose to remake himself into a hero for scientific research liberated from Christianity, southern inferiority, and the stifling orthodoxy of academia. Gabon served Garner as a site from where he could spin tales that displayed his racial and intellectual superiority. No longer was he merely an obscure figure visiting zoos; he could demonstrate his heroism and then use it to bring attention to his theories of animal communication. Atlantic travel furnished Garner the space to reinvent himself.

Perceptions of masculinity threatened the very promise of a new glamorous life through selfless research in supposedly dangerous locales. It is hard to determine if his unusual ideas or his supposedly unscrupulous fail-

ings as a man posed a greater impediment to his career. Critics derided
Garner as a man who did not have the self-control to properly conduct
scientific research and as a cowardly liar. These gendered critiques put into
question his veracity and manliness, perhaps even more than his lack of
formal education did. When wealthy philanthropists and influential scientists did decide to put their trust in the Virginian traveler after 1910,
Garner's self-sacrifice and daring won them over, not his science. Such a
victory proved to be a fleeting one, as he would be judged after the mid-
1920s by his theoretical positions rather than by his Gabonese adventures.

Garner's career also indicates how his masculine persona required sacrifices. Instead of celebrating his southern background like many New
South intellectuals, Garner rode the rails to northern cities and left his
thoughts on the Lost Cause and segregation unpublished. He made no effort to attract the aid of universities or scientific institutions in the South.
Science and African travel allowed him new venues to prove his self-
worth without reference to his background. His family also had little place
in his ideal of individual heroism abroad. Although his melodramatic relationship with Ida Vera Simonton and his pleas to his son to settle down
indicate his deep-felt belief in personal honor and male authority over
women, he fell short of his own standards over and over. Safari hunters
like Theodore Roosevelt could go home again, and American Protestant
missionaries could promote their models of domesticity in their work, but
Garner never managed to reconcile his public career with his views on the
proper behavior of a gentleman.

Although Garner's name has never received attention as a southern
public intellectual, his career suggests a rethinking of Appalachian and
southern history in several ways. First, southern freethinkers and advocates of scientific research did not necessarily endorse common views on
Christianity and southern identity in the Gilded Age and Progressive Era,
even as they wrestled with similar problems and experiences with others—a point that often seems lost in the literature on southern masculinity. Also, southern expatriates like Garner reshaped tropical adventure
and Atlantic networks to promote their own personal interests as well as
to participate in common portrayals of Africa as a backward, dangerous
place where white men could put their claims of superiority to the test.
American accounts of life in African colonies might disseminate similar
ideas of race and gender, but they also could express individual ambitions

that varied wildly. Finally, Garner represents an example of a southerner who chose to downplay the Civil War and uphold the ideal so praised among northern middle-class communities: rugged individualism at the expense of kin and regional identity. Appalachian historians need to bring men like Garner into their discussions, particularly as their experiences show how townspeople were willing to distance themselves from their rural neighbors and associates, as well as from the heritage of disappointment and economic hardship that came after the Civil War. Garner may have taken a unique course, but he shared many of the same anxieties of other white men in the New South.

CHAPTER FOUR

Race, Knowledge, and Colonialism in Garner's African Writings

IN 1908 RICHARD GARNER PONDERED racial conflict in America from an extremely unlikely vantage point: the shores of the Fernan Vaz Lagoon on the southern coast of Gabon, part of the French colonial empire in central Africa. As a former Confederate soldier from a slave-owning family, Garner had a very intimate perspective on the "Negro Question." Few proponents of systematic discrimination and the legal disenfranchisement of African Americans made their way to Africa. Yet Garner joined the chorus of white southern intellectuals and politicians in the Gilded Age and Progressive Era who drew on their supposed expertise to teach northerners the value and necessity of Jim Crow laws. Garner could simply have stayed home in southwestern Virginia to pen his conventional attitudes toward race. "Sooner or later in the course of human progress mankind will learn by experience the egregious and irreparable error of trying to develop the negro race along the same lines of mental and moral evolution as the white man. . . . The negro race will remain for ages if not forever, a social dead-weight that will necessarily handicap the white man without benefitting the negro," he declared in a pessimistic evaluation of the future of race relations in the United States.[1] Although Garner's outlook did not make him distinctive, his experiences in Gabon made him unique.

How Garner employed his long sojourns in southern Gabon to critique colonial rule and promote himself as an expert on racial issues are the central topics of this chapter. Garner's ethnographic research and his claims to have mastered the languages of simians gave him the tools to give evidence on behalf of the need for social and political segregation. Like so many other enterprising southern white intellectuals in the Progressive Era, Garner aimed his essays at northern whites rather than limit his audience to southerners. The reconciliation of white America after its shattering during the Civil War could extend abroad. Just as some supporters of the Spanish-American War presented the conflict as a means of showing the strength of a truly united (and white) nation, Garner saw Gabon as a place where he could find evidence to support the Jim Crow South. Furthermore, Garner's contention of unbridgeable mental and physical differences between blacks and whites also could be used to justify aggression against people of African descent. Yet Garner also found French rule in Gabon to be a corrupt and brutal enterprise that ultimately degraded Africans. Although he ultimately chose to quiet his attacks on French rule, his writings demonstrate how white southern constructions of race could be used to critique European colonialism.

Garner's writings offer an important contribution to the growing literature on U.S. empire and on American constructions of Africa. Scholars have observed for decades how American representations of Africa in the Progressive Era supported white supremacy at home. However, African travel narratives by white southerners have not received much attention. Most treatments of American portrayals of Africans in the early twentieth century have focused on popular texts like Tarzan novels or well-publicized events such as Theodore Roosevelt's safari to British East Africa. National considerations of race took precedence over regional concerns.[2] Garner consciously placed his discussions of Gabonese society and colonialism into a dialogue with the American South's legacies of slavery and Reconstruction. He thus broke entirely from Charles Darwin, whose commitment to abolitionism led him to conceive of humanity as a single species made up of many races. By contrast, Garner tried to give credence to the long-standing dream of southern politicians to deport African Americans across the Atlantic. However, Garner did not agree with famed Thomas Dixon's efforts to prove to northern readers that white southerners could furnish valuable lessons for imperial expansion.[3] Atlantic networks of knowledge could be put to work to support Jim Crow at home, thus showing how the color bar was built on transnational dialogues.[4]

Garner joined a long tradition of anti-imperialist racism that contended colonial rule foundered on the inherent racial divisions between whites and other peoples.[5] Uses of Africa to undermine American imperialism have not received much attention, but more well-known figures such as Mark Twain and University of Chicago anthropology professor Frederick Starr articulated similar opinions to Garner in their reviews of Leopold II of Belgium's Independent State of the Congo. While Anglo-American advocates of colonial reform, such as the Congo Reform Association, have recently been valorized as pioneers in international human rights work, many of its members shared with Garner a deep disdain of cultural hybridity and intimate relationships that crossed racial boundaries. Garner might not have been well known, but he provided a white southern tone to the criticism of U.S. and European empire. He also was quite conscious of the legacy of Atlantic slavery.

Garner's story has value as an example of an aspiring public intellectual's efforts to become a respected expert on racial matters in Progressive Era America, especially for an individual whose ideas deviated greatly from other white southerners. He despised the rhetorical flourishes of Anglo-Saxon pride, the Democratic Party, and Christianity. Such unorthodox views, as well as his lack of institutional backing, placed Garner at a severe disadvantage in promoting his program. Becoming a racial expert not only offered an opportunity to join what Nina Silber has called the "romance of reunion" between white southerners and northerners but also allowed Garner to enter the mainstream of segregationist thought.[6] Furthermore, Garner's determination to belong to an international community of biologists gave him the chance to offer typical white middle-class southern views on race to a wide audience. Although the memory of Garner quickly vanished after his death, between 1892 and 1920 he had become a nationally known sensation for his theories. Even if Garner's goals vastly exceeded his ability to promote his own ideas, his writings offer an unusual chance to see anti-imperialist racism from the perspective of a white southerner in Africa.

Race in Garner's America

Before Garner first set foot in Gabon in 1892, he had developed a range of ideas about racial relations. Although few of his personal writings and stories from his life before becoming a public intellectual have survived, his account of his childhood indicates that he had close friendships with several older black slaves.[7] "Uncle" Charles, a slave preacher, became Garner's childhood idol, and Garner even considered becoming a Protestant minister because of Charles's kindness and gift for telling stories. He also turned to an African-born slave known for his skills in making herbal remedies. These relationships are invariably presented as friendly. The only hint that white dominance ever was challenged in Garner's childhood was a very brief reference to a public execution of a black man.

Yet Garner's surviving accounts of his early life and adolescence are marked by conspicuous silence about racial conflict, particularly once the Civil War began. Although southwestern Virginia and eastern Tennessee are often considered to have been bastions of loyalty to the federal government, middle-class townspeople from railroad towns like Abingdon gravitated toward secession. Garner's father and several of his brothers served in the Confederate army as well. They had profited from the construction of railroads through the region in the 1850s, and an increasing number of slaves worked in businesses, mines, and farms in the decade before the Civil War.[8] Violence was commonly used to demonstrate the authority of slave owners. Once the war began in earnest, anxieties about rebellious slaves spread in southwest Virginia, especially after two slaves charged with killing a white man were executed in 1862.[9]

Garner himself had an intimate knowledge of violence against African Americans. He may have participated in a massacre of black troops. By late 1862 and early 1863, Abingdon became a center of Confederate operation in a region that had become a battleground.[10] Local Confederate and Federal supporters battled one another in a series of bloody guerilla campaigns. Garner himself joined up with the Third Tennessee Mounted Infantry on 22 June 1864, a unit that reformed to battle Union forces in east Tennessee and southwest Virginia after its defeat at Vicksburg. Union troops led forays into southwest Virginia, and a series of battles pit the Third Tennessee against Federal troops in the summer and fall of 1864.

The United States Fifth Colored Cavalry Regiment engaged Confederate troops on 1–2 October at Saltville, an important mining center not far from Abingdon.[11] After a Federal assault on the mine's defense was repulsed, Confederate troops allegedly massacred black troops the day after they had surrendered. Garner's Third Tennessee Regiment arrived on the scene at the waning moments of the battle. Although historians remain divided over how many African Americans died, it is clear that some black prisoners were executed. In a 1916 letter, Garner recalled serving in the battle but did not discuss its bloody aftermath. It is unlikely that Garner himself participated in the killings, but it seems inconceivable that he would not be aware of what happened. Abingdon was devastated by a Federal raid in December 1864. During this attack, Garner was taken prisoner nearby in the town of Paperville, near Bristol, Tennessee.

The personal and national defeats Garner experienced during the final years of the Civil War and the tumult of the Reconstruction era very rarely entered his discussions of racial difference. He mentioned the Saltville incident only briefly to a colleague and did not make reference to the executions of prisoners. Garner wrote a long poem in 1889, "Feast of Fire," that lauded the Lost Case, but in his later work hardly ever mentioned his fighting days. His family's own challenges after the war also remained obscure in his later writings.[12] Garner attended the private high school Jefferson Academy from 1865 to 1867. After finishing his secondary education, Garner traveled west rather than remain in Abingdon, which had been partially burned to the ground in December 1864. Census records indicate he lived in Texas in 1870 but then made his way back to Appalachia. A news report contended that he fought against Apache communities.

Why did Garner not mention these experiences of dislocation and upheaval in his writings on race? One possible explanation lies in Garner's self-representation as a caring patriarch of animals and Africans. Naturally, cold-blooding killings and the collapse of slavery would hardly reflect well on such a public persona. Garner's rosy memories of his childhood, with simple and loyal African Americans respecting white authority, could provide comfort to a man who had seen his hometown partially destroyed and the order of slavery vanquished. Several fictional stories set in the antebellum era featured docile people of color and generous masters. In one tale, an aging Native American in Louisiana, the last of his tribe, earned the favor of nearby whites by not trying to cross "the barrier of race distinction," as his "ready acquiescence in that law of society gave him all the more prestige." A kindly planter agreed to follow the aged man's dying request to keep his faithful dog, which Garner deemed to be typical, because "some of the most beautiful and some of the most pathetic stories of mutual affection and fidelity [are] between men and dogs among the American Indians."[13] Another tale, "Caesar's Ghost," describes the friendship between a simple and loyal old slave to a drunken master. When Caesar the bondsman passes away, his alcoholic owner becomes convinced that his slave's ghost returns to comfort him. The master's wife decides to trick her spouse by dressing as a ghostly Caesar and then advises the inebriated plantation owner to quit drinking, which he promptly does.[14] Such a comic tale presents Caesar as a happy and faithful servant.

Like so many other white southerners in the Gilded Age and Progressive Era, Garner portrayed slavery as the best means of improving people of African descent. Slavery harnessed their economic potential instead of offering an education not suited for their intellectual capabilities. In 1905 he wrote his longest piece of race relations, observing, "During a century and a half of slavery [the Negro] was taught the arts of civilization and during that period he applied them with more diligence than he has since liberation. . . . As a whole, he was more civilized and productive as a slave than he has been as a freeman." Emancipation and access to education had not helped matters. Garner stated, "That literary education has failed to elevate the negro no longer admits of debate, as the criminal records of all civilized countries testify. That the moral status of the negro is lower under the white man's civilization than it is under his own, no one who is familiar with the native life of Africa, can, for a moment, doubt."[15]

Garner's racial theories rarely wavered from the dominant themes of immutable racial differences and the validity of Jim Crow. He contended time and again that the development of African American communities since emancipation had borne out his thesis. Unprincipled white agitators had lured black people with impossible promises of social and intellectual equality: "The negro of his own accord never demanded nor aspired to social or political recognition among white men until fanatics of the latter race sought him as a candidate for such favors." People of African descent in the United States had become dependent on white generosity and now loitered "as a beggar at the gate of Dives, whining about his rights and licking the boots of charity." In brazen defiance of the historical record, Garner added that black people had received as much equal pay and economic opportunity as any white person.[16] Such stereotypes of needy, lazy black people and kindhearted whites may have had little to do with the actual events of the Reconstruction period, but they certainly allowed Garner to ignore the challenge to white southern male dominance that took place in the late 1860s and 1870s.

Even before Garner first went to central Africa, he already publicly expressed some of his views regarding race in print. Written in 1892, his first book, *The Speech of Monkeys*, contains numerous descriptions of his experiments with zoo animals, which demonstrate his disdain for people of color. In one of his accounts, he pretended to beat a black girl with a stick to entertain a monkey. After he "sacrificed her to the altar of science," he then "drove the child from the yard with a great show of violence." Not only did Garner indulge in what he deemed to be playful aggression toward African American children, but he also argued that his theories of animal language could be applied to distinctions between Caucasians and other peoples. For example, Garner asserted the greater complexity of European languages in comparison with African languages. "It is said of some of the African tribes that their gestures are more eloquent than their speech," he noted. He also claimed that two different species of monkeys "differ in mental caliber as widely as the Caucasian differs from the Negro."[17] To explain the distinction between the common chimpanzee and a supposedly rare variation known to southern Gabonese people as the *kulu-kamba*, Garner opined, "they are the white man and the Negro of a common stock."[18] For the Virginian researcher, differences among animals ran parallel to racial hierarchies among humans.

Garner did deviate from more typical southern configurations of race in one important way. The very word Anglo-Saxon appalled him. He wrote his son that he would prefer to "hand over rule to the American Negro" than cede power to foreign immigrants and that the United States had left the Anglo-Saxon world behind at Yorktown. "We are not an Anglo-Saxon people — we are an American people, and stand for American civilization," Garner announced.[19] A poem dedicated to Americans speaking in English accents, "Caucasian Monkeys," stated this practice "is simply a thing to deride, revolting to all true American pride."[20] This view was hardly typical in an age when Anglo-Saxon unity was the clarion call of southern Democratic politicians, although Garner's measured support for

Republicans in nation politics was a common view in the mountain South at the turn of the twentieth century.

Other than this one point, Garner's prejudices did not distinguish him from many other white southerners. On one hand, his experience living in Gabon with Africans for more than two decades made him a quite different kind of racial expert than most other writers on racial issues, save for a handful of Protestant missionaries. Southerners had long used purportedly secondary ethnographic sources and popular stereotypes about Africa to justify white supremacy. Governor and senator of Mississippi in the early twentieth century, James Vardaman, a legendary advocate of black disenfranchisement, declared in a notorious speech in Congress that the African American had "never created for himself any civilization" and that "when left to himself he has universally gone back to the barbarism of the jungle."[21] Armchair anthropologists such as the sociologist Joseph Alexander Tillinghast could provide comments such as the "psychic nature of the West African exhibits most of those immaturities so common among uncultured savages," but they never set foot in Africa.[22] Instead, they relied on slave traders' accounts, colonial ethnographies, and travel narratives such as those from the English traveler and amateur ethnographer Mary Kingsley. Only southern racial theorist Earnest Sevier Cox, a future proponent of Nazism and eugenics, could boast of having visited Africa repeatedly between the 1890s and 1914. Garner thus provided a synthesis of observations about Africans and African Americans to justify political disenfranchisement and the deportation of black Americans to Africa.

Garner as a New South Ethnographer

Garner compared his observations of African Americans with Gabonese societies to develop a set of racial characteristics that he claimed existed among all black people. His methodology was far less systematic than his academically trained counterparts. He steered clear of the juxtapositions of physical features and objects that furnished data for physical anthropologists in the late nineteenth and early twentieth century. Descriptions of social and cultural practices from various communities all over the African continent also did not enter his discussion, as they commonly did for many armchair anthropologists before the 1920s. Despite asserting the similarities between Africans and African Americans, he made little reference to specific cultural practices that could be used to show commonalities along racial lines. Instead, Garner confined his attention to an examination of intellectual ability, the ability to apprehend abstract concepts, and personality traits. His firsthand knowledge of southern Gabonese societies made him stand out from others.

In keeping with other southern intellectuals seeking to convince northern white readers of the necessity of segregation, Garner presented himself as a credible and evenhanded witness. Garner's transatlantic background made him an unusual figure. "I have had such a vast experience with the negro in all conditions of life and under such circumstances that I am fairly familiar with his native habit and nature. . . . I am not prejudiced against him and few people have done more good in proportion to his means than I have done," he noted in one unpublished essay.[23] He did not belong to the rival camps of hateful enemies of black people nor the supposedly deluded believers of absolute racial equality. The conclusion of Garner's account of a Gabonese canoe paddler's belief that the moon sometimes hid itself in the ocean led Garner to note, "Instead of regarding [the man] as an incorrigible heathen and wishing to sweep him off the earth, I sympathise [*sic*] with his innocent philosophy."[24] This posture of the knowledgeable and kindhearted friend also appears in one of Garner's few published works on Gabonese societies. In a 1902 article on cultural practices in Gabon, Garner declared he preferred African beliefs to those of white people and noted, "It is absolutely painful to any student of primitive customs to witness the contempt in which the majority of white men hold the native institutions of Africa."[25] Garner had extended the popular metaphor of caring masters and white patriarchs protecting African Americans from dangerous ideas to encompass Gabonese people as well.

Racial differences had engendered different forms of human intelligence and personality traits, according to Garner's theories. Some of his claims corresponded with other common portrayals of cowardly black people. After years of conflict with Gabonese neighbors and workers over pay and land, Garner bitterly declared that "Negro loyalty" was a myth. Black people acted only out of fear.[26] Yet Garner refused to categorically declare in print that Africans were less intelligent than white people and instead believed that people of African descent had a radically different kind of reasoning from white people. In fact, he found the "diabolical cunning" of Gabonese people far more threatening than their ability to use violence against him.[27] What separated black from white was the ability to understand abstract concepts. "The negro is essentially a materialist and his mind only deals with material things. . . . Spirit apart from matter is as unthinkable to the negro mind as the immensity of space," Garner stated.[28]

Evidence for this idea could be drawn from closely observing southern Gabonese spiritual beliefs and practices; Gabonese people had no concept of deities at all. Sacrifices of goods and alcohol offered to *ombwiri*, powerful spirits associated with particular locations, did not demonstrate reverence for a god. "All alleged sacrifices . . . are simply the payment of tribute for a specific act. . . . There is nothing sacred or reverential in the act or the thing appeased," he wrote shortly before his death in 1920.[29] Gabonese people honored only the forces they feared. Garner recalled one journey when his close ally Anjanga, a clan chief of the Nkomi people living on the Fernan Vaz Lagoon in southern Gabon, prayed to a storm to calm itself on behalf of Garner and all the Nkomi paddling Garner's canoe. On the same trip, Anjanga and his crew commanded their American employer to keep silent as they passed over waters known to be inhabited by Milamba, a mighty *ombwiri* that became angry at the slightest noise. Garner concluded, "There [*sic*] supplications are as really prayer as those of more intelligent races of mankind who address them to some deity supposed to be supernal if not supreme."[30]

Gabonese spirituality may not have been factually true, but even for a skeptic, it held value. Garner pointed out that indigenous religious beliefs maintained social order and corresponded with the intellectual capacity of Africans. Yassi, a masked spirit who entered Galwa clan villages by the Ogooué River to punish criminals, was a scourge to Catholic and Protestant missionaries. Garner felt that Yassi merited praise and that his power truly came from his communal nature: "a creature of the people, as represented in the society which creates him, . . . [Yassi] is simply a clever piece of detective work, well planned and well executed." The psychological fear of physical harm from poison ordeals, instead of Yassi's supposed mystical power, made him a force to be reckoned with. Garner argued that Gabonese spiritual beliefs demonstrated the creativity of Africans, as well as their mistake of attributing their accurate observations of natural phenomena to imaginary causes. "The wise-men of Africa . . . are aware that the odours inhaled by certain plants produce sleep and that others produce insomnia. . . . They do not know how these things are accomplished, but they are conscious that such is the fact. The secret of the witch is knowing the plants that produce certain effects. . . . This is the sum and essence of witchcraft," he noted in 1902.[31] He repeated the same argument two decades later.[32]

One reason for Garner's unwillingness to simply condemn African intellect as unfailingly flawed came from his need for African knowledge. Without hunters and traders to inform Garner about animals, he had little ability to conduct research, save for the direct observation of animals. His recognition of the value of Gabonese knowledge — even as he carefully obscured the individual identities of most of his African associates from view — placed him in an awkward position for a proponent of white rule in the United States. Yet this respect for indigenous information did not lead Garner to accept Africans as equals. Instead, Garner argued that only sharp boundaries between Africans and Westerners could ensure mutual respect.

For all of Garner's professed tolerance and understanding, he often made clear his sense of white superiority. Complaints about African workers constituted one of the main topics in his diary from 1905 and 1906. "These lazy dirty beggars resort to anything on earth to avoid work," ran one typical quote.[33] When several wives of a Nkomi clan leader fled to Garner to escape their husband, the American bemoaned the situation. Gabonese women sometimes turned to paternalist missionaries in Fernan Vaz to leave their spouses, but Garner was not up to the task. "It is rather an unpleasant duty to have to be the 'Father' of a lot of savages whose social horizon is but little above that of the gorilla or chimpanzee and such is my task among these wooly cannibals," he wrote.[34] The Yassi secret society showed how "in mental stature these simple [Gabonese] people are but little children."[35] Garner loved to show off the intellectual inferiority of Africans by pretending to have supernatural powers. Phonographs became a means of speaking to the dead, and a porter unlucky enough to put some acid from Garner's packages in his hair was informed that he was a victim of Garner's power to harm thieves through magic.[36]

Garner put an anthropological spin on his sense of superiority. When Garner donated some ethnographic artifacts to the Smithsonian in 1900, he provided a "fetish tusk" imbued with supernatural power that he had purchased from a Fang-speaking community on the Como River in northern Gabon several years before. The elephant tusk was filled with a variety of objects, including several hairs purportedly taken from Europeans. This powerful item, supposedly the "most potent known fetish known among the tribes of Central Africa" furnished its possessor with "the sagacity of the white man with his gun powder and *savvy*."[37] It furnished convenient evidence that Gabonese people recognized white intellectual superiority and power, even if Garner's limited knowledge of Gabonese languages suggests such a description reflected his own opinion rather than that of his anonymous informants.

Garner could cross racially demarcated cultural frontiers without suffering any damage, but Africans who traversed the same dividing lines became dangerous. When Nkomi people held healing ceremonies for Garner, he thanked them for their hospitality even as he assured his readers that he considered the entire ritual to be based on mere superstition.[38] Garner contended that his excursions into Gabonese cultural practices left him unscathed from any hint of physical or psychological degeneration, in marked contrast to common anxieties about Westerners living in colonial locales who did not maintain physical and social segregation. Africans were not so fortunate, particularly converts to Christianity. Cultural hybridity threatened the purity of white and African races, and Christianity was among the biggest causes of African degeneration. "We always knew when Sunday came in the bush," he assured his American readers in 1920, "because on that day our boys refused to work, having learned from the missionaries that Sunday was a holy day of idleness."[39] The missionary education of Africans was "simply a comic opera feature that makes the negro a subject of ridicule without benefiting any human being. . . . A native who can read or write, no matter how poorly, will no longer do any kind of manual labor, he becomes meddlesome, officious, and presumptious [*sic*] to an overwhelming degree."[40] Garner wrote down coastal Gabonese versions of Christ's resurrection that described the Holy Spirit as the master of powerful supernatural forces that made it invisible. A retelling of Samson's story by some Fang-speaking converts featured a mystical talisman made of elephant bone and made a dispute over marriage payments the central reason for Samson's fight against the Philistines.[41]

Why did Garner find missionaries so appalling? First, they defied his claims of intellectual difference. By offering Christianity to Africans, they foolishly lured people to cross natural racial boundaries: "the missionary not only loses his own prestige with the convert but sacrifices that of own race; and in return gains nothing, except the fawning of the sycophant, who is like an elephant in swaddling clothes." Africans who held themselves "with genuine dignity" remained deaf to the siren charms of Western education and churches.[42] Such views might seem at odds with dominant views of southern white Christians, but Garner himself was a steadfast atheist who considered Christianity to be "a propaganda of falsehood, a nursery of hypocrisy and night-mare [*sic*] of human progress."[43] While American Protestant and French Catholic missionaries expressed misgivings about the ways Gabonese Christians might challenge racial hierarchies through dress, writing, and other means, they did not question evangelization. For the Virginian freethinker, however, Christianity was a menace to the natural separation of the races. Hypocritical abolitionists had duped black people in America with empty promises of the "false light" of social equality, and missionaries did the same in Africa. The end result was a path into "the treacherous quicksands of social despair."[44] Such distrust for Christianity would gain adherents later in the twentieth century, with racial theorists such as Madison Grant and Lothar Stoddard, but hardly any southerners took aim on Christianity as a threat to the color bar. Even eugenics advocates used Christendom as shorthand for white unity, which would have revolted Garner.[45]

Racially distinct forms of intelligence and cultural practices strongly informed Garner's views on colonialism. American expansion outside of North America never emerged in his writings, but Garner presented himself in the first decade of the twentieth century as a fierce critic of French rule in Gabon. "If the French republic is conizant [*sic*] of the evils that lurk under her flag and does not openly rebuke and suppress them, she is unworthy of her place and prestige among nations; and if ignorant of them, it is time that the mask that hides her colonial scurvy should be removed," he fulminated in one essay.[46] Like other Americans who opposed colonialism before World War I, Garner drew on fears of cultural hybridity, political corruption, and degeneration in his condemnation of the French government. Garner thus presented himself as a heroic protector of African freedom and distanced himself from the tyranny of French rule, even as he benefited from the colonial administration and tried to appropriate some of its power for himself. An American southern intellectual could thus correct the excesses and errors of French imperialism in the same manner as he enlightened northern whites about racial verities.

A Confederate against French Tyranny

Garner's long stays in Fernan Vaz allowed him to witness the grim results of the concessionary company system, and his critiques joined together his ideal of separate racial development with a concern for personal liberty. The undated essay "The French Congo" constituted Garner's fiercest and most elaborate critique of French colonialism and the CCFV. Written between 1904 and 1909, it lists the daily indignities that Gabonese people endured. For Garner, slavery appeared better than the miserable status of Africans in Fernan Vaz: "As a slave he would have only one master to serve and conciliate [*sic*], but as a citizen of this free colony, he has to endure the whim of many masters, who not only seize his lands and products with recompense, but [in] addition make him pay tribute to his alien oppressors." He compared the prosperity of Gisir-speaking villages in the southern Gabonese interior during his initial visits before 1900 with their present wretched state under the CCFV. Now, "no work or trade of any consequence was going on, the country was stripped of goats, fowls and every living kind of domestic animals, plantations had been devastated . . . and the whole country seemed to be on the verge of famine." Empty villages with still-burning cooking fires greeted his arrival in many cases. He learned from a few Gisir people brave enough to meet him that "seeing a canoe approach with a white man wearing a helmet the native supposed it to be a tax hunter with his gang" and fled with all the property they could carry (2, 4).

People had good reason to evade guards and administrators. One Gisir leader told Garner through an interpreter that the local administrator had forced him to collect rubber for an undetermined time, because younger men had run away to avoid paying taxes. Although he had collected an amount of rubber worth more than what he owed to the state, the official maintained that the Gisir chief had paid only half of his tax. At another settlement, Garner observed a man pulling a small goat to a canoe while several women cried in anguish. It was the last goat in the village, and now it had to be handed over to the administrator to pay the tax of 5 francs. The government considered the goat to be worth only 1.5 francs, so he still owed 3.5 francs and expected to be imprisoned for his inability to pay (4–5). A government caravan collecting taxes had "a half dozen carriers loaded like pack mules" with booty confiscated from Gabonese communities unable to pay in hard currency. Their load included chickens, iron pots, drums, native knives, food, cloth, and a monkey for good measure (6). Women had their cloth ripped away from them by guards. At the town of Otombi, some guards had bound the men, "outraged the women," and grabbed all the plantains, mats, pots, and other valuable property in the town; they even snatched pins used to decorate women's hair off their heads. When the Gisir chief tried to complain, the French administrator had him imprisoned at the coastal town of Ombouè, on the testimony of his troops (8). When French officials met the arrival of troops guarding prisoners and requisitioned merchandise and animals, guards kept a fair amount of goods for themselves.

Just as concessionary companies never paid close to market value for natural resources, administrators and their African auxiliaries commonly undervalued items and animals collected to pay taxes. French francs hardly could be found anywhere in Fernan Vaz: "These grasping [concessionary companies] pay next to nothing for the native products and but a small part of that in cash" (10). Class shaped Garner's critiques even more than those of better-known advocates of reform in colonial Africa, such as E. D. Morel. "Many of [the officials] are drawn from the menial ranks of civilization. . . . In their own estimation, they are Napoleons and Wellingtons and Blackstones and no doubt imagine that their names will live in history as empire builders and throne makers, instead of cooks and stewards and bush traders," he asserted. Garner declared, "No individual white man has ever given willingly an hour of his service or a penny of his wealth gratuitously to the cause of civilization, sordid gain is the fundamental motive of all their actions."[47]

The proclivity of French officials for their intimate relations with Gabonese women also angered Garner, who set himself apart from most European men in the colony for his supposed disdain for cross-cultural sexual encounters. Mission-educated coastal Gabonese women had become famous in West African circles for their willingness to become the sexual partners of Europeans. Historian Rachel Jean-Baptiste has noted how some coastal women preferred temporary relationships with European men because of the abuse and lack of opportunity they sometimes suffered in marriages with local men.[48] Outside of Libreville and other mission centers, male chiefs had offered female sexual partners to visiting Europeans since the late eighteenth century. Such relationships utterly revolted Garner: "White men hobnob with native chiefs and magnates on terms of fraternity when by doing so they can make slaves of their sons and concubines of their daughters. . . . No penalty is inflicted for such trivial offences [as rape], although they are committed time and again by the very agentcies [*sic*] of civilization whose duty is to prevent and punish them" (14). Garner's public abhorrence of interracial sexual contact—notwithstanding his private collection of photographs of Gabonese women—corresponded with constructions of bourgeois white masculinity in Virginia after the Civil War, but only European and American missionaries held similar views in Gabon.

Garner's critiques of Gabon under the concessionary firms resembled those of more prominent critics of colonial rule and corruption at home, especially those presenting themselves as crusaders who truly represented civilized virtue. E. D. Morel made similar arguments in 1903, as he described how French concessionary companies confiscated British traders' merchandise in Gabon: "A flagrant instance of the general topsy-turveydom, the prostitution of law and order, the inversion of parts, the perversion of facts, [and] the general organized hypocrisies of which the French Congo had become the scene."[49] Individuals opposed to the annexation of the Philippines in the United States contended that the Anglo-Filipino war degraded democratic ideals and threatened the supposed racial homogeneity of America.[50] Muckraker journalists eviscerated politicians as enemies of freedom for their willingness to accept bribes and rig elections. Garner too could present himself as a heroic investigator of trusts in Africa, as well as someone ready to reveal the horrors of imperialism.

Garner's solution for French colonialism called up the specter of an impossible though cherished dream of generations of southern white intellectuals and politicians. African Americans and Gabonese could join together and live in a free rural settlement in an independent Gabon, where they could work effectively on their own, far from whites. There, a black American could there be free of "vain dreams of social and political beatitude" that came with a Western education, creating an "ornamental consumer of other people's products . . . sitting around with a book in his hand and an empty stomach." Garner recommended southern Gabon for its sparse population, rich potential for agriculture, and good harbors at Cape Lopez and Loango. One slight obstacle for Garner's theory lay in the fact that the French government controlled this area. Gabon "can never be developed by [the French government's] methods and for a small consideration, no doubt they would decamp and allow a colony of real home-seekers to develop it," he concluded optimistically. Perhaps Switzerland could be trusted with the task of supervising the new nation.[51] Such a plan for a Gabonese Liberia defied reality and lay dormant in Garner's private papers. However, U.S. race experts and politicians kept calling for emigration from the 1880s through the 1930s.[52] Southern plans of African American emigration could save Gabon.

Nothing came of this plan, but another failed Garner venture put together racism and anti-imperialism. Garner's bizarre 1906 plot to place New York socialite Ida Vera Simonton as the queen of rebel Mitsogo clans at war with the French administration also tied together critiques of French colonialism with white superiority. Once Simonton took the (entirely imaginary) throne, she then could negotiate an end to the concessionary system in Gabon in return for an end to the rebellion.[53] Gabonese people would be under the benevolent tutelage of Garner, the southern race expert, and his female American partner. The French government would learn the value of reform and credit Garner's extensive knowledge of southern Gabonese societies for solving the long-standing Mitsogo revolt. By positing the French as crude exploiters, Garner also maintained his own supposedly impartial position on Africans.

His published writings reveal none of the impassioned calls for liberty. Even his essays on race never appear to have reached a large public audience, save for the few articles on Gabonese spiritual practices already mentioned. Because racial questions remained paramount in his notes, it is hard to determine why they never made it to print, except perhaps for a general lack of interest among editors. It does appear that he did discuss race in his public lectures in New York City, Washington, and other cities, but unfortunately only his fragmented notes and unlabeled lantern slides have survived.[54] Why Garner kept relatively quiet about French rule is easy to explain: he needed the colonial government. In a magazine article published in 1920, written only several months before Garner's death, the American researcher repeatedly praised French officials for the aid they provided to his final expedition to Gabon. Governor of Gabon Casimir Guyon had sent Garner's request for aid to his superior, Governor General Gabriel Angoulvant of French Equatorial Africa. The governor general consented to all of Garner's wishes. The American received special permission to hunt without any restrictions, did not have to pay customs duties on ammunitions and supplies, and could expect support from all administrators in Gabon. "One must live in Central Africa in order to know what toadies the petty officials are, and to be able to appreciate the value of these kind orders," Garner observed. Later in the article, Garner assured his readers of the organization and benevolence of French colonialism in Gabon.[55]

One should not assume that conditions had improved for people living in Fernan Vaz in the fifteen or so years between the magazine essay and Garner's unpublished critiques. The CCFV company sputtered into oblivion because of the 1913 collapse of world rubber prices and World War I. Although free trade returned by 1920, the war proved to be an unmitigated disaster for Gabonese people. Thus, Garner's abandonment of his earlier opposition to colonial rule did not come from improved conditions for Gabonese people, but rather from the advantages Garner had obtained from the French government. The American researcher would have had enormous difficulties outside of the aegis of French officials, no matter how often he portrayed himself as an independent and intrepid entrepreneur. Privileges extended to Garner as a Western resident of Gabon — military support, the right to carry modern firearms, and authorization to trade in an area where concessionary companies held an absolute monopoly over exports — allowed him to survive. The loss of his favored status would have forced Garner to leave the colony. Thus, Garner had every reason not to embarrass the regime that kept him in business. The advantages of colonial rule's support for white supremacy outweighed the risks that could have come from a more vocal condemnation of French authority. In the end, Garner's brand of anti-imperialist racism found common cause with French racist justifications for empire.

Conclusion

Garner's personal views on race and colonialism reflected his southern background and his unorthodox views on religion. Rather than dwelling on his virulent dislike for many Gabonese people and his own family's legacy of slavery, he instead shaped a persona of a detached scientific observer of racial issues. He aspired to become a famed expert on simian language and a specialist on race issues through his African experiences. His eccentric views on religion made his own views on race somewhat at odds with many white southerners. Yet his ostensibly scientific foundation for believing in separate racial development would herald the rise of scientific racism. The Social Darwinist visions of Madison Grant, Lothar Stoddard, and other American scientific racists followed much of Garner's platform. Other southern race specialists put together negative depictions of African society to justify white supremacy in the United States, but the little-known Garner had far more information at his disposal than more prominent writers, and his approach rested on his Atlantic crossings.

Garner argued that slavery furnished white southerners with a proper understanding of race that eluded Europeans less directly linked to Atlantic bondage. French officials were incompetent and rapacious jokers, and they served as foils against which Garner could reprise his self-representation of white southern paternalism. Anti-imperialist racist ideas could thus teach French administrators the errors of unchecked exploitation, instead of the supposedly more benign form of white rule in the United States. Catholic missionaries likewise were as duplicitous as northern abolitionists in defying the eternal truths of racial difference. And as always, Garner made himself out to be an authority who sought out only the best interests of Africans. In his own way, Garner became a transnational theorist of race relations through his time in Gabon and the U.S. South, even as he eschewed more popular considerations of the color bar as a global issue.

Just as he avoided discussing the collapse of slavery and the undermining of the social order he had known as a child, his public discussions of race never hinted at the precariousness of his situation as a relatively impoverished white settler in southern Gabon. Garner's pose as a well-experienced observer of Africans and African Americans allowed him to imagine himself as a wise patriarch looking after the true needs of black people on both sides of the Middle Passage. The necessity of Jim Crow could be proven by evidence drawn from his early life and his later career in Gabon. While some racial theorists of the American fin de siècle sanctioned violence against African Americans to uphold white power, Garner did not raise the questions of lynching or violent disenfranchisement. Instead, Garner's brand of white supremacy cloaked itself in patronizing charity. This position also allowed Garner to ignore the use of force involved in maintaining white privilege, whether in southern Appalachia or central Africa. His conception of an antebellum era of strict racial hierarchies and his disdain for colonial ventures thus made him a vocal member of the U.S. anti-imperialist racist tradition. Just like most anti-imperialists who raised racist objections to American empire, Garner reconciled himself to the French domination of Gabon. Colonial privileges proved too instrumental for Garner's ambitions to reject them, just as his own assertions of compassion for people of color were built on a foundation of coercion. Although an Atlantic African colony and America had quite distinct configurations of white supremacy, many whites concurred with Garner's choice to downplay the inherent violence of colonial domination in Africa and racial disenfranchisement in the United States.

CHAPTER FIVE

African Animals for White Supremacy

NO ONE CAN DENY THAT RICHARD LYNCH GARNER loved animals, no matter how little he cared for people. Living creatures fascinated him even as a boy in southwest Virginia. "Among the blue hills and crystal waters of the Appalachian Mountains, remote from the artificialities of the great cities . . . nature was the earliest teacher of my childhood, and domestic animals were among my first companions," he recalled in the opening pages of his 1900 book *Apes and Monkeys*.[1] Besides presenting himself as a son of the countryside attuned to natural phenomena, he expressed here a growing sentiment among many Americans in the late nineteenth and early twentieth century. Katherine Grier, Harriet Ritvo, and others have noted how middle-class Americans and Europeans depicted domesticated animals as objects of emotional attachment and key participants in family life.[2] Tamed animals acted as entertaining servants worthy of love and proper guidance, and their owners could compare their care of their furry friends favorably to the rough treatment dogs, cats, and other animals experienced in working-class communities.

Of course, Garner did not merely befriend animals already familiar to American households as good pets. Throughout his career as a public intellectual, he also sought to convince his readers and listeners that the denizens of the Gabonese rainforest could be as touching and worthy of love as lapdogs and Persian cats. Garner spent decades battling against the belief that human beings were unique among the entire animal kingdom because of their intelligence or their capacity for emotion.

As an atheist, he also discarded the belief that God had separated man from other animals. While such ideas were radical in Progressive Era America, Garner's portrayal of African animals as loving and entertaining resonated with broad cultural trends regarding the place of animals in U.S. society.

The rapid proliferation of zoological gardens in the United States after the Civil War allowed visitors to observe elephants, monkeys, and other animals firsthand. Newspaper writers and zoo operators alike worked tirelessly to make these animals into lovable and innocent creatures that could draw crowds. Bronx Zoo curators such as William Hornaday tried to sway visitors to the cause of conservation through these endearing animals. Nigel Rothfels has rightly noted that Hornaday "underscored a narrative of the civilizing and urban bourgeois ambitions for the world — through the educated and philanthropic activities of the best members of human society, the rest of the world's people and all the world's animals could find shelter in benevolent care." Similarly, British zoos also relied on individual exotic animals like Jumbo the elephant to bring in financial support and display their concern for animals.[3]

Garner made common cause with his erstwhile patrons to promote a new vision of tender and comic animals, but his exhibitions also held great value for his own individual agenda. He strove to portray his relationship with primates and dogs as precious and important to him as his links to his own human family, although Garner could demonstrate the validity of his kindly patriarchal persona far more with chimpanzees and faithful canines than in his unsuccessful marriage. Animals also could serve as devoted servants Garner could dress, name, and control in similar fashion to the slaves of his childhood in Virginia in the 1850s. Garner rejected sharp distinctions between the emotional and intellectual capacities of humans and other primates. He thus had no difficulty making comparisons between Africans and domesticated animals, in which his pets proved far more respectful of white dominance than his Gabonese workers and neighbors.

Garner's treatment of animals could demonstrate racial superiority over people of color. Dorothee Brantz has contended in nineteenth-century Europe that "the treatment of animals was often viewed as a way to measure the standard of civilization."[4] While African animals became kindly friends, African people were considered spiteful and capricious. This distinction was a harbinger of representations of African fauna later in the twentieth century, as colonial governments and Western public audiences increasingly depicted animals as vulnerable creatures who could be saved from African people only through the wise supervision of foreigners.

Besides exposing the changing place of African animals in the individual and collective American imagination during the Progressive Era, Garner's discussion of animals also furnishes a rare glimpse at Atlantic variations in human-animal relations. Gabonese and European residents of Gabon had their own ideas about animals that rarely matched Garner's American ideals of pet keeping. European men used African people and environments as a foil to highlight their individual and racial capacity for action.[5] On the southern African frontier in the age of high imperialism, European descriptions and treatment of dogs often served as a means of social demarcation between settlers and local people. Settlers and officials justified the taxation and eradication of local dogs by presenting these animals as nuisances and threats to European property, while praising their own pets as faithful companions in comparison to Africans.[6] Despite this growing literature on canines and empire, the ubiquitous presence of domesticated monkeys, apes, and other pets has drawn very little attention. English, French, and German settlers enjoyed showing off their exotic animals as well as their more familiar pets.

Pet ownership of primates in Gabon diverged significantly from the uses of domesticated animals in Japan, Europe, and North America. While Garner's uses of animals ran parallel to general trends in North America and Europe, foreigners working in Gabon did not seek to emulate models of family life and civility that held sway in their homelands. The treatment of chimpanzees and gorillas in settlers' hands often mirrored the brutality and derision Europeans in central Africa imposed on colonial subjects. At the same time, European and American narratives of primates also fit with Kathleen Kete's recent questioning of the vogue of pet keeping in nineteenth-century Europe: "Is not pet keeping, then, another way to hide from ourselves the real violence between humans and animals beneath an image of sensibility, or even as a means to deflect us from awareness of the violence between ourselves and others in an age of class conflict and global domination?"[7] At least in Gabon, Europeans and Americans actually did more than use loyal animals as a foil against Africans; they sometimes savored the violence of colonial conquest through pet ownership.

Garner and His Monkeys

Garner had developed decidedly unusual opinions regarding primates and other animals well before he first reached Africa. From 1884 through 1891 Garner regularly observed different species of monkeys in zoos in Cincinnati, New York, and Philadelphia. He concluded that these animals shared many personality traits with humans. One monkey lay in the arms of a zookeeper: "She would put her little slender arms about his neck and cuddle her head up like an injured child. She would caress him by licking his cheeks and chattering to him in a voice full of sympathy, with an air of affection worthy of a human being." Garner argued her tales "must have been of the deepest woe" and that "she was begging her keeper not to leave her alone in that great iron prison with all those big bad monkeys that were so cruel to her." Her cries were "real tears, and are doubtless the result of the same causes that move the human eyes to tears."[8] Not only did monkeys and apes employ a form of communication analogous to human language, but they also shared similar emotions to humankind. Garner warned his readers not to assume human superiority: "There are, perhaps, instances in which the mind of an ape excels that of man by reason of its adaption to certain conditions. It is not a safe and infallible guide to measure all things by the standard of man's opinion of himself."[9] Although Garner did not make reference to the rise of humane societies and antivivisection leagues in Europe and North America, he shared a common concern with these organizations about the emotional well-being of animals.

Garner's views on primate emotions became even more pronounced after his first Gabonese sojourn in 1892 and 1893. His book *Apes and Monkeys* describes a jungle idyll in which the American researcher had become a patriarch to a band of amusing and simple chimpanzees and gorillas. Moses, an emaciated young male found by a Gabonese hunter near the Ogooué River, became the first member of Garner's attempts to form a family of tamed chimpanzees. It ate with Garner at his table, slept in a house Garner built for his new pet, and ripped apart newspaper tablecloths after it finished its food (120–121). Despite such failings, Garner praised Moses for his emotional bond: "I know of nothing in the way of affection or loyalty that can exceed the affection of my Moses" (134). "Tears of pity and regret" rolled down the researcher's face when he re-

turned from a short trip and discovered that the animal had become ill
(141). After Moses died, the American wrote, "To say that I did not love
him would be to confess myself an ingrate and unworthy of my race" (142).

Garner contrasted his love for chimpanzees with what he deemed to be
the callousness of Gabonese people. When Garner later kept chimpanzees
together in a cage in Libreville, he noted how Gabonese spectators teased
the animals: "The ruling impulse of nearly all natives appears to be cruelty,
and they cannot resist the temptation to tease and torture anything that
is not able to retaliate" (157). After he acquired his chimpanzee Aaron,
Garner described how "cruel hunters" killed Aaron's mother; he had tried
to defend her body against her "slayers" (145). Such scenes illustrated
Garner's own sensibilities as a caring white man. In 1894 a Gabonese servant complained about Aaron's penchant for grabbing passing branches
as he carried the chimpanzee on his back. The chimpanzee refused to eat
food given by the Gabonese worker and tried regularly to bite his unlucky
handler. "So the quarrel went on until we reached the [Rembo Nkomi]
river," Garner noted, "but by that time each of them had imbibed a hatred for the other that nothing in the future allayed" (148–149). By presenting Aaron and the Gabonese man as rivals equal in their capricious-
ness, Garner denied a sharp distinction existed between Africans and
chimpanzees.

Garner conflated his views on women with female primates. When
Moses died soon after Aaron's arrival, the American researcher traveled
from Fernan Vaz to Libreville to purchase a mate for his remaining pet.
His new conquest, Elisheba, reminded him of "certain women that I have
seen that had soured on the world. She was treacherous, ungrateful, and
cruel in every thought and act. . . . She was utterly devoid of affection"
(154). Garner projected his ferocious antipathy toward New Womanhood
on the monkey. When Garner's plans to convince New York City socialite
Ida Vera Simonton to become his literary and romantic partner failed in
1907, he described Simonton in exactly the same manner as Elisheba: sul-
len, deceitful, and heartless.[10] Yet the animal's illness in a Liverpool zoo
made her a tragic figure, and her companion Aaron's concern for the dying Elisheba was so moving that "even the brawny men who work about
the [zoo] paused to watch him in his tender offices for her, and her staid
keeper was moved to pity by his kindness and patience" (170). Chivalrous
male chimpanzees could watch over vulnerable female animals, just as

Garner portrayed himself as a courteous guardian of female animals and human women.

Garner's later unpublished work continued in this vein. In a short autobiographical sketch written around 1908, he noted his lack of success in maintaining friendships and gaining the public acclaim that he so craved. "Conscious of my isolation, I have long since turned to nature for that companionship that responds which no fellow creature of my race had occasion to accord me and in the great kingdom of the untamed I have found my friends," he concluded.[11] In 1906 a Boston women's magazine published a description by Garner of his daily routine in his home in the southern Gabonese rainforest.[12] His idyll of jungle domesticity displayed a paternal concern for animals. Frolicking monkeys played "like happy little Brownies" and mischievously stole fruit from Garner's gardens. Sugar cane and pineapples planted near the house lured chimpanzees from the forest so that Garner could observe them. In the household, Garner described his West African cook Baba as his lieutenant, followed by his pet chimpanzee Plato. Tellingly, Baba's Gabonese wife ranked beneath Plato and the one chicken on the premises. Lessons and play with Plato constituted the main pleasure for the lonesome scientist, and with his primate companion, he ate the meals prepared by Baba. In the account animals received far more attention than any African, especially for the joy that they provided observers: "Not an hour spent in the jungle that doesn't reveal some secret of nature hitherto undreamed of." The pristine wilderness evoked a sense of grandeur unbound by human constraints, and his mastery of its secrets was a source of pride.[13]

An incomplete essay that Garner wrote on two orphaned monkeys highlighted Garner's views on animal emotions, as well as his own role as a guardian of innocent creatures. In a rare moment of self-reflection, he noted, "No doubt human sympathies are wasted on animals, because we view their afflictions in the light of our own afflictions and ascribe to them deeper passions than they really have . . . but they are the kind of wealth anyone can afford to squander."[14] That being said, love for animals offered its own benefits for humans, even if such care led to the death of many creatures. Like many collectors, Garner left the dirty work of capturing most animals he purchased to Africans rather than accept his own part in killing. Two Gabonese hunters brought Garner one tiny monkey after having caught and eaten its mother. It chirped with pleasure once

Garner held it, and the researcher believed it wanted to tell him how cruel the hunters had acted. Moved by pity, he then bought the monkey.[15] The orphan now had a parent, and Garner named this surrogate child Dinkie. After teaching it to eat, he then carried Dinkie around in a sling. A month later, a slave owned by one of Garner's Nkomi neighbors brought him a tiny monkey barely bigger than Dinkie. The young man had found the animal while clearing Garner's garden. A 1906 newspaper article by Garner described the two as "saucy imps" who always wished to be around their master.[16]

When Africans emerge in Garner's accounts of animals, they usually acted as cold-hearted killers in contrast to the American's kindness or as inadvertent obstacles to proper research. In August 1908 Garner lamented how his West African cook set a dog on some chimpanzees that had come to eat sugarcane from the field specially designed to attract primates to the settlement.[17] These accounts deflect both Garner's role in the deaths of animals and the social and political context in which he lived. Hunters from various Gabonese communities came to hawk live and dead animals at Garner's settlement nearly every day, according to Garner's diary.[18] And as discussed elsewhere, Garner's ability to flout the concessionary company monopoly over trade made him a very popular customer. To bring the American specimens, Gabonese men often killed animals. Whether by passing visitors or by a slave hired out from a nearby Nkomi master, Garner again relied on partnerships with local people to maintain his household's grounds and to bolster his animal collection. Far from being isolated, Garner's entire venture of forming an animal family relied on African knowledge and labor. Rather than admit the fallacy of his imagined, self-made world, Garner placed his attention almost entirely on his personal bond with animals.

As was the case with so many other American and European animal collectors in colonial Africa, Garner eliminated most references to the Africans who actually brought him the animals he domesticated. In 1910, photographs of his pet chimpanzee Susie invariably accompanied articles in the popular press about his triumphant return to America, but instead of showing Susie's rainforest past or her African handlers, they clearly underscored Garner's parental care. The chimpanzee now also belonged to a domesticated fantasy of childhood in which animals deserved the same loving treatment as human children. A children's magazine featured Susie in a dress sitting at a table with blocks and a black doll.[19] Another article had an image of Garner sitting with his pet on the same chair, while another photograph displayed a white girl and Susie sharing blocks.[20] Susie handed spectators toys and balls at Garner's command.[21] An essay published shortly after Garner's death in 1920 showed a picture of Susie in Garner's embrace, and its caption noted how the animal agreed to sit for the camera only after its owner calmed it in its own language.[22] This imaginary world where pet ownership could bring humans and primates together was bolstered by Garner's admission that a sign of the elevated nature of Susie's behavior was its ability to treat other animals as pets.[23] In these portrayals of Susie, the African context of transnational exotic animal trading fell by the wayside. Across the Atlantic, though, Westerners did show off their pets in ways that did not correspond so well to typical early twentieth-century American understandings of pet ownership.

Domesticated Primates and Monkeys in Colonial Gabon

Missionaries, colonial officials, and travelers rarely discussed their pets in their own writings, but the animals emerge frequently in Garner's essays. When Garner brought Elisheba and Aaron aboard a ship headed for Europe, he met a sailor who also had a male chimpanzee. The two men decided to allow Elisheba to choose between her two suitors. "It was the drama of 'love's young dream' in real life, in which every man, at some period of his young career, has played," Garner noted (165). Other European acquaintances of Garner also celebrated the antics of their chimpanzee pets. Otto Handmann, the German consul in Gabon in the 1890s, allowed his chimpanzee to sit at the table, drink with his guests, and join in applauses for toasts. A German trader's female chimpanzee became addicted to beer (182–184). Another chimpanzee, owned by a steamer captain in Gabon, could untie any knot made by the crew, scrubbed the deck with a broom, and knew how to work a corkscrew (203–207). Other European traders simply kept chimpanzees chained to trees day and night (185–186).

Why did Europeans own these pets in Gabon? It is striking how little primate ownership appears in missionary and official correspondence, but Garner's writings illustrate how commonly foreign residents of Gabon enjoyed showing off their primate pets. To answer this question, one must first investigate changing understandings of pet ownership in Europe and North America. The increasingly sentimental relationship between humans and certain species of domesticated animals was part of broader trends that celebrated emotional ties and constructed home as a private, feminine space.[24] The overwhelmingly male American and European population in Gabon before World War I hardly fit with any tender ideal of family life. Instead, European men celebrated heavy drinking, sexual relationships with African women, the regular use of violence against Gabonese workers, and hunting.[25]

Based on Garner's writings, English and German chimpanzee owners clearly enjoyed the mischief caused by their animals. European owners of these animals showed off both their animals' ability to mimic human activity and their inability to do so consistently. Instead of praising their animal companions as loving servants, owners derived pleasure from the imperfect resemblance to themselves. Such mixed pleasures held particular salience in a colonial African context in the late nineteenth century, as European and American residents of the colony bemoaned the danger of cultural hybridity of Gabonese people.[26] Coastal Gabonese people portrayed themselves as a community that had accepted Christianity, knew European languages and cultural practices extremely well, and deserved full political equality with French citizens. Administrators and missionaries critiqued such claims by presenting cultural hybridity as a source of degeneration, much like their counterparts in India.[27]

Jeremy Prestholdt's recent examination of consumption practices on the East African island of Zanzibar offers a series of insights relevant to European consumption of chimpanzees and gorillas. In Zanzibar both European slave redeemers and slave owners sought to rename, redress, and literally remake captives as objects to be displayed and enjoyed. Through names and dress, missionaries and slave masters alike presented themselves as kind patrons seeking to improve the lives of their wards. Owning animals also allowed Europeans to imagine themselves as caring masters who spared chimpanzees and gorillas from the brutality of African hunters. The irony that the European demand for primates may have very well increased the number of animals killed did not disrupt this fantasy. Prestholdt contends that Europeans deployed narratives of East African consumption of European goods to construct a coherent European modernity that Africans could not successfully imitate.[28] Chimpanzees followed suit. They could undo but not make knots, eagerly drank alcohol but could not stop themselves from drinking, and swept with brooms without actually cleaning. Because many Europeans in Gabon blurred the distinction between apes and Africans, monkeys and chimpanzees' antics could offer commentary on the supposed inability of Africans to appropriate European social and cultural practices.

Unfortunately, Garner spent much less time describing how Gabonese people treated domestic animals. Dogs were a ubiquitous part of village life throughout the colony, as hunters relied on canines to track down and chase game. At least based on folktales collected by the noted Gabonese priest and scholar André Raponda Walker, dogs were seen as wild and looked down on for their inability to control their ravenous hunger.[29] Monkeys would be occasionally tamed as well. Garner offered one tantalizing case of a domesticated chimpanzee he inspected in a Gisir-speaking clan settlement in southern Gabon. A Gabonese man reported that some people had captured the animal when it was very young, and it had become a companion to all of the children in the town. On command, the chimpanzee fetched water in a gourd and guided people out of their homes to see Garner. When the American tried to buy it, he discovered the price was double the amount that was normally paid for a slave in southern Gabon (177–182). Several older men in a limited sample of twenty people interviewed in the Fernan Vaz region in December 2008 recalled how chimpanzees were tamed in the past. "They were like dogs. You just had to give [chimpanzees] some bananas. After a few weeks, they would stay around the village and beg for food," recalled Jean-Robert Ngomba, an approximately eighty-year-old village chief.[30] Unfortunately, such intriguing references do not furnish much evidence on primates domesticated by Africans in the late nineteenth century.

It seems clear enough that African practices of animal domestication had little impact on how Europeans treated and discussed their own pets. Garner's search for a kind of domesticity that transcended differences between species may have had little to do with drunken chimpanzees stumbling about trading posts and steamers, but in each case Westerners employed tamed creatures to demarcate the distance between themselves and Africans. Garner's primate family was full of love, even as it was surrounded by vicious Africans. German, British, and French traders could enjoy their pets' antics as a form of entertainment and social commentary.

Perhaps the most vivid example of the racialized nature of Garner's pet stories was his long paean to his faithful hound Bubu. Written in 1908, this forty-page essay exposed how pets stood at the frontiers between Garner and his African neighbors.

A Thin Strand of Whiteness: Dogs and White Supremacy in Fernan Vaz

Garner's considerations of his own dog reveal how dogs entered into struggles between settlers and Gabonese people as well as in disputes among Africans. Because Garner chose to live far from officials and missionaries, he was surrounded by potentially hostile and demanding Gabonese neighbors. Race greatly shaped Garner's depiction of Bubu as an exceptionally intelligent and earnest ally in a world populated by emotionally distant and unscrupulous Africans. In characteristic fashion, Garner made claims designed to bolster the value of his ethnographic and intellectual capital: "While I frankly confess my attachment to my dog, I shall try to record his virtues with as little bias as possible. . . . I believe my habits of studying the psychic qualities of animals . . . allow me to do this with little or no partiality." Such expertise led him to believe that Bubu's single white hair indicated "patrician blood," namely of European fox-terrier stock. Though a mongrel, as were "all the dogs of European descent bred in Africa," Bubu's manners "alone give evidence of cultured heredity and his innate intelligence indicates a well-bred ancestry."[31] Dog breeding gave rise to similar discussions of racial hierarchies, from Japan to South Africa to England.[32]

In a colonial Gabonese context, such rhetoric plays off the single most popular topic of discussion by Anglo-American and French visitors in the late nineteenth and early twentieth century: the hazards and pleasures of cross-cultural contact. Coastal Omyènè women had long developed a range of sexual liaisons with European men. Omyènè free people loudly asserted their sense of equality to Europeans based in part on their intimate familiarity with European cultural practices, such as following fashions and chatting with English tourist Mary Kingsley about London art galleries. Garner, reflecting his own performed identity as a white upperclass southern gentleman, himself regarded with horror how commonly traders and officials openly lived and slept with Gabonese women. Yet, despite his canine friend's mixed heritage, Garner deemed him a loyal European. The American's sense of racial ancestry uncovered the thin strand of his whiteness in his dog, thus negating his usual sense of dismay at anything that hinted at cultural hybridity.

Not only did Garner believe he and Bubu shared a common origin in Europe; they both disdained most Africans. Many of the anecdotes that attest to Bubu's cunning demonstrate the dog's determination to guard what Garner deemed to be his property: "When natives bring anything to sell, the dog permits them to put the commodity down . . . but once [the African has] released its hold upon it he must not touch it again, unless my cook or myself deliver it to him . . . and as for anything that [Bubu] knows belongs to [Garner's] place, he permits no stranger, except for a white man, to do so. Any white man, however, that has so far visited me, he permits to take hold of everything, and has never growled at one."[33] Grievances about theft by domestic servants and neighboring African settlements fill Garner's surviving diary. Garner deemed himself to be the sole owner of the land he lived on.

Bubu supported Garner's claim to own land. Garner contended that he had definitively bought the land around his house, but Nkomi people regularly demanded payment from him and held local leaders responsible for protecting their guest. What Nkomi people viewed as legitimate rights to collect natural resources around Garner's home was viewed by the American as theft. Itutu, an old Nkomi man, regularly collected pepper growing near the American's settlement. One day, Bubu saw Itutu take the pepper and then "sprang at his hand and took the leaf [containing the pepper] away from him" (17). Garner's dog also bit the leg of another Gabonese man who had received Garner's permission to cut sugarcane on his concession. The dog apparently did not understand that the man was not stealing. Supposedly, Bubu "never forgets a theft that he witnesses . . . and I am now satisfied from his conduct towards certain other natives, that he has seen them do something that he believes that they had no right to do" (18). Bubu permitted Garner to deny local people their rights over land and property and imagine himself as the true owner of the land around his home, with a devoted animal servant to protect his property, instead of as a guest subject to his Nkomi landlords.

Even as Garner sought to keep himself and his pet free of the demands of Africans, his praise of Bubu also allowed him to cope with other anxieties about sexuality. Garner strongly identified with his dog in part because both had been rebuffed by the same woman: Ida Vera Simonton. The wounds of Garner's failed relationship with Simonton are on display throughout the essay, although he carefully avoided naming her. Bubu became the only tangible benefit Garner actually obtained through his ties to Simonton. He first met the dog when she brought him to Garner's home, where the dog "really did not seem to know whom he ought to obey" (5). Simonton's lack of empathy and her fickle attentions were hazards to Bubu. "His owner took but little interest in him except to spasmodically pet and caress him once or twice a day" (6). When Simonton chose to leave Fernan Vaz for the small port station of Cape Lopez in the spring of 1907, Garner and the dog accompanied her. After they arrived, she moved into the local administrator's house, leaving them both behind. Bubu came up to Garner "with a rush as though he was overwhelmed with joy at seeing me again and at the same time acted as though he had come to induce me to go back with him to find his mistress" (8). Simonton wanted nothing to do with the dog: "He ran to her and reared himself upon her as if to express his pleasure and exact her caresses. Instead of caressing him, however, she pushed him away" (10). Bubu eventually started following Garner around the small settlement. Although Garner's cook returned Bubu several times to Simonton, he made his way back to his new friend (10–13). At last, his former owner boarded a steamer without Bubu.

Garner presented Bubu as a victim of the unwomanly and uncaring Simonton. Bubu became a canine victim of an American New Woman abroad. The bohemian and theatrical circles Simonton frequented in turn-of-the-century New York often evoked fears of deviancy. By contrast, Bubu's capacity for affection earned Garner's praise. Bubu also displayed his sense of abandonment. "There can be no reasonable doubt that the heart of this hapless little dog was stung and bleeding at the unexpected and unkind treatment of one whom he loved with all the honesty of his race and the one of all others from whom he had the right to expect the right of a kind return," Garner noted (14). Icy-hearted suffragettes thus lacked the inner sense of caring that separated them from well-bred dogs and gentleman adventurers.

This depiction of a white man's alliance with a slightly white dog against Africans and rogue American women is undermined by the somewhat uncomfortable presence of Garner's African employees. Much of

Garner's knowledge of his loyal subject came from his cook from Sierra Leone, as this servant had to watch over Garner's pet. This West African immigrant held a prominent role in many of Garner's essays. A fair number of English-speaking men from Sierra Leone had moved to coastal Gabon as early as the 1860s, brought by English traders and American missionaries.[34] Garner could not bring himself to treat the cook as a valued collaborator. By placing Bubu and the cook on the same plane in his dog stories, Garner followed a common practice in southern African dog tales that considered European pets and Africans as equals in intelligence and behavior.[35]

The Sierra Leonean immigrant seems to have taken care of the dog as much as Garner himself. He watched over the kitchen, where Bubu occasionally killed chickens or stole eggs, and had been assigned to bring Garner's canine companion back to Simonton (10–13). In turn, Bubu exempted the cook from the hostility he showed other Africans (17–19). Although the American suggested that he had sworn himself to sharing some food with Bubu "as long as I had a crust [to eat]" and boasted that Bubu always "accepted with apparent gratitude" fare from his table, it was the cook who actually prepared the meals (13, 15). African workers thus performed the tasks that constituted Garner's kindness to animals in Gabon.

Garner described Bubu in ways that demeaned the intelligence of his cook. He downplayed his cook's role in aiding his efforts and even recorded several of the cook's requests for pay in broken English as examples of the failure of African intellects to master European languages.[36] The Sierra Leone migrant and his employer had a long debate over whether or not the dog had a habit of killing chickens and stealing eggs from the kitchen (21–25). When Bubu walked up to Garner or the cook with a chick in his mouth, his owner refused to punish him on the ground there was no proof he meant to eat it. "The circumstantial evidence was sufficient to convict any dog of sucking eggs, a thing for which thousands of dogs have suffered unmerciful punishment and many have suffered death as the penalty for the crime, but the cook had not seen Bubu take the egg from the nest" (24).

Again, Garner's emotional sensibilities came to the fore to rescue his noble canine servant from Africans. The cook caught Bubu with an egg again and argued with Garner that the dog had stolen it from the chickens. "Instead of flogging the dog I caressed him and gave him every assurance of my approbation," Garner wrote. While Garner interpreted Bubu's gesture as a misguided expression of loyalty, his assessment of his cook's suspicions was less kind. Garner concluded, "The cook reluctantly admitted the indictment against little Bubu was not sustained by the proof but I think he halfway regretted the dog's innocence, for it is one of the singular mental characteristics of the negro race to believe the worst of every living creature and want every suspect condemned on general principles" (25). The American's use of reason to demolish the prejudiced opinions of his African servant demonstrated his supposed superiority.

A final vignette from the "Bubu" essay denotes the crueler side of Garner's animal kingdom. In colonial Gabonese villages, Europeans and Americans commonly ate chickens donated by community leaders as a sign of hospitality honoring their arrival as valued guests. After Garner reached one village, a chief ordered a group of children to grab one bird. The dog's response brought out a particularly colonial moment of humor for Garner: "He nipped four of [the children] and every one in the same part of his anatomy, just where the native African finds a sore to be the most inconvenient. I had to hold the dog while the chickens were caught" (34). This story reveals much about Garner's conceptions of African masculinity. His dog violated the boys' manhood and rejected the chief's generosity, just as Garner felt he was not obligated to respect local men's rights over his patronage. And, as the comment about disease suggests, Garner viewed African men as debased. Venereal diseases were endemic in Gabon during the early colonial period, and American and European observers often depicted Gabonese women as both physically ill and morally corrupt.[37] The comment also fits well with Johannes Fabian's analysis of humor in German ethnographic accounts, in that humor designed to mock Africans served as a means of maintaining social distance and as a coping mechanism: "European humor is safe behind an enclosure that marks resistance to the pull that Africa exercised on the minds of explorers."[38] Dogs and jokes both guarded the borders of Garner's interactions with Gabonese people.

Garner's piece shows how pets in colonies might act as sentinels of racial and gendered hierarchies in the eyes of their owners. By guarding property rights and disrespecting indigenous claims of social reciprocity, Bubu was an important asset for the isolated and vastly outnumbered Garner. In comparison, the American considered Africans deceitful, ignorant, sexually deviant, and cruel toward humans and animals alike. Considerations of animals also articulated individual concerns as well as expressing broader issues of social distance, racial theories, and the anxieties of living in colonial Africa. Finally, Bubu illustrates how Garner could enjoy the spectacle of suffering Africans as much as any other Westerner living in Gabon. For all of his bluster about his empathy toward people of African descent, his uplifting of human-animal relations offered little place for his Gabonese neighbors.

Conclusion

Richard Lynch Garner may have become a household name in the United States in the early twentieth century for his theories of animal language, but he also promoted an ideal of white pet keeping that fell in line with dominant understandings of human-animal relations in this period. In American popular thought, apes had long symbolized racial difference and the savage nature of African life. Garner countered this idea by celebrating the similarities between humans, monkeys, and apes. Not only could monkeys speak a language, but they could suffer loss and feel love as well. Happy chimpanzees playing with toys and watched over by Garner prepared the way for the anthropomorphic approach of exotic animals in American zoos and museums. Gentle white keepers followed Garner's model of enlightened animal care, even if Garner himself was consigned to the forgotten ranks of pseudoscientific cranks by the 1930s. As was so often the case in the transatlantic circulation of knowledge, African perspectives on animals did not reach North American audiences.

African animals did not always represent African savagery in need of male white conquerors, but cuddly chimpanzees and loyal dogs could still articulate the anxieties and the aspirations of American racists. While other scientists in the Progressive Era refused to accept the idea that people of color could contribute valuable knowledge, Garner denied that black people had the emotional capacity to love animals. Apes and dogs could detect the innate truth of the color bar. Conservation ideology and practice in the early twentieth-century United States and in European colonial possessions implicitly expressed similar ideas of the noble Western treatment of animals and environments in comparison to the supposed cruelty and irrationality of indigenous peoples' uses of land and animals. Similar ideas surfaced again in postcolonial Africa, particularly with the advent of what Dan Brockington has named "fortress conservation," in which foreign environmental organizations treat African hunting and land use as inherently a threat to endangered animals, with little concern for land rights or the actual impact of African practices. A century after Garner's stay in Gabon, ecologist Michael Fay traveled through some of the same terrain that Garner had visited and promoted an image of wilderness menaced by encroaching Africans.[39] Unambiguous references to racial difference no longer came into play, but the pathological view of African interactions with the environment juxtaposed with enlightened Westerners remained.

Like other Westerners in colonial locales, Garner also enforced his ideal of social and physical segregation from Africans through animals. Apes and monkeys allowed Garner and other Western residents of Gabon to distance themselves from Africans, whether by mocking the ability of Africans to claim equal status with Europeans or by suggesting that animals were more worthy of love than colonial subjects. Bubu's aggression toward Gabonese people became a mark of his superiority, as the canine enforced Garner's claims to land and authority that went against landlord-stranger conventions long used by coastal Gabonese communities to hold Western guests accountable to local people. Emotional ties between pets and privileged foreigners served to gloss over and even justify the violence and exploitation that undergirded colonial rule. Portrayals of animal-human ties thus became tools to construct and maintain white supremacy on both sides of the Atlantic.

Richard Garner. All photographs are from Photographs folder, box 5, Richard Lynch Garner Papers, National Anthropological Archives, Smithsonian Institute, Washington, D.C.

Richard Garner with rifle. Like many Westerners in colonial Africa, Garner sought to display his mastery over African people and animals through hunting. However, Garner had to rely on Gabonese guides for help with capturing and killing animals.

Southern Gabonese leader with entourage, circa 1900. Nkomi, Gisir, and other Gabonese leaders controlled access to the labor and knowledge Garner needed in southern Gabon.

Official reception, perhaps in Gabon. Garner claimed to be a firm opponent to French colonial rule, as it supposedly threatened the strict cultural separation of the races that Garner espoused. Gabonese communities sometimes turned to Garner to mediate disputes with the French government. However, the American researcher relied on French permission to hunt and capture primates.

Three Gabonese women. As a white southerner, Garner claimed to despise French officials and British traders who had intimate relationships with African women, but his collection of photographs of seminude Gabonese women suggests his personal life may have not always matched his segregationist rhetoric.

Two Gabonese diviners. Although as an atheist Garner derided all religion as superstition, he underwent indigenous healing rituals, and sometimes as a negotiating tool, he presented himself to Gabonese people as a supernaturally gifted figure.

Garner with dog and chimpanzee aboard ship. While Garner's theories of primate languages found few adherents among the American scientific community, his ability to bring chimpanzees and gorillas alive from Gabon to the United States earned him respect and patronage from the Bronx Zoo and the Smithsonian Institution.

Richard Garner introducing his Gabonese chimpanzee Susie to a child, 1910. For Garner, emotions and languages existed among primates as well as humans. As an atheist who felt that humans were merely another animal species, Garner wished to convince the general public how chimpanzees could speak and care for others. In these discussions, Garner valorized primates at the expense of Gabonese people.

CHAPTER SIX

An American Sorcerer in Colonial Gabon

AROUND 1906 RICHARD LYNCH GARNER FACED a typical problem during his long stay on the Fernan Vaz Lagoon on the southern Gabonese coast. Garner had long experienced traveling along the rivers and shores that connected the lagoon to both southern inland Gabon and the important trade routes along the Ogooué River farther to the north. During one trip, the American followed a caravan of Nkomi workers from Fernan Vaz to the small administrative and trade center of Cape Lopez, located on an island facing the Ogooué River Delta.[1] He traveled with an armed convoy assigned by the colonial administration to repair and reconstruct the telegraph line that stretched from Cape Lopez to Fernan Vaz. In the early twentieth century, the small French colonial establishment at Cape Lopez acted as the headquarters of the limited French presence on the southern Gabonese coast. According to Garner, these escorts had great value: "By this means I could stop over on any place en route without having to pay and support an idle caravan while I was resting."[2] Despite his own bitter dislike of French colonial policy, Garner relied on French administrative efforts to use technology and armed force to better control Gabon.

Close to Cape Lopez, six new carriers from the village were drafted to join Garner and the Gabonese leader of the caravan. Garner remarked that the caravan leader chose to take the lightest burden: a toolbox. As the journey commenced, Garner realized he had forgotten to pack a bottle of carbolic acid and hurriedly placed it in his toolbox. A sudden downpour after a few hours sent Garner scurrying for cover. He noticed that the caravan leader seemed nervous when he asked about the box. The man asked Garner, "Master, you no get dem witch medicine?" When Garner then asked if the leader had stolen something, he denied it but said in pidgin English, "Well, dem rain he be *mbuiri*; he be proper witch rain. He get fire." *Mbuiri*, a word Garner drew from his imperfect knowledge of the Omyènè language used as a lingua franca in southern Gabon, referred to supernatural forces.[3] Garner responded, "That's no witch palaver — that's the devil after you. You've been doing something bad and he's getting close to you."[4] The leader promptly panicked and begged Garner for a cure, while the other carriers blamed the toolbox for the mysterious trouble, especially when the leader tried to make someone else carry it. The group continued to march to Cape Lopez.

At the end of the journey, the leader again begged Garner for help against the *mbuiri* set against him. Garner fell asleep instead of settling the issue but discovered the cause of the Gabonese man's anxiety the next morning. A strong odor of carbolic acid awoke Garner. When he came to the house where Garner was staying, the caravan leader confirmed that he indeed had stolen it. Almost all of his hair had burned off. The man had thought it was lavender to be used for hair care and had suffered as a result. Garner gave him an ointment for the wounds, and the man asked for pardon while blaming the *mbuiri* force in the rain for his troubles.

Garner pointed to this story as a "comic opera" that exemplified his own intellectual superiority over Africans, who were in his view "the most superstitious of all mankind."[5] For the American observer, humor served as a means of maintaining social distance from Africans and separating himself and the reader from empathizing with pain. The use of parody and derision were key elements to constructing ethnographic narratives that celebrated white supremacy.[6] The use of technological advances as a means of justifying Western dominance was part and parcel of European and North American colonial expansion.[7] There is no question that Garner's ethnocentric focus and his less-than-fluent command of local languages and cultural practices placed limits on the accuracy of his descriptions and his analysis. However, it would be a mistake to dismiss the value of Garner's accounts for understanding political and social changes in southern Gabon.

Not every Gabonese associate of Garner could so easily fit into his hierarchies of simple-minded Africans and civilized whites. An Omyènè-speaking coastal Gabonese leader living on Lake Onanguè, located close to the Ogooué River, sent Garner a letter in 1907: "I inform you that you belong to me in this country, since it is me who is responsible for this place. According to our customs, each family has its own land. Thus you must know that I will come chat with you as soon as I arrive."[8] While southern Gabonese societies might be included on Western maps as part of the colony of Gabon, Gabonese leaders still asserted their right to act as patrons of foreign visitors like Garner. The American became embroiled in disputes over what his reciprocal relationships with African landlords actually required of him. On other occasions, Gabonese people forced Garner to enter disputes with European trading companies.

To settle disagreements, Garner and Nkomi people in southern Gabon used a common if contested vocabulary of supernatural forces and political rights between indigenous people and foreigners that had emerged in the age of Atlantic slavery. Garner presented himself as a victorious apostle of science as he manipulated phonographs and chemicals to ensure local people would accept his claim of absolute freedom from local demands. While Nkomi men treated Garner as a privileged guest that needed to pay homage and tribute to his local hosts in similar fashion to slave dealers and European trading firms in the past, the American contended he could selectively respect or reject reciprocal obligations placed on him by his Gabonese neighbors. The limited presence of French authorities in southern Gabon before World War I and Garner's own marginal position in a French colony forced the American adventurer to show deference and publicly endorse the supernatural beliefs of his hosts. In these negotiations, Garner often presented himself as a herald of modernity who awed Africans with Western technology. Southern Gabonese people placed his new objects into a long tradition of commercial exchange of mobile power objects that were believed to contain mystical forces. In indigenous spiritual beliefs, Garner's objects entered local frameworks of supernatural power in similar fashion as other new innovations. Missionaries were also said to command great spiritual forces.

This chapter offers an important contribution to the historiography of the early colonial period in equatorial Africa in several ways. Historians of francophone central Africa have long debated the ability of the co-

lonial state to remake African societies in the brutal period of concessionary firms between the 1890s and the 1920s. Georges Dupré, Catherine
Coquery-Vidrovitch, and John Cinnamon have documented the destruction and social dislocation brought about by military occupation and the
rapacious policies of concessionary firms in Gabon and other central
African colonies.[9] Jan Vansina has gone so far as to declare that equatorial
political traditions collapsed by the 1920s due to the violence that came
with the initial decades of European occupation. Jane Guyer and Samuel
Eno Belinga likewise argue that the ability of equatorial African leaders
to control and develop access to individuals who had specialized knowledge also came undone in the maelstrom of early colonial rule. In 1981,
in the only published monograph written on Fernan Vaz, anthropologist
François Gaulme argued that the French state completely supplanted the
precolonial Nkomi monarchy between 1898 and 1912.[10] A group of scholars
in the 1990s highlighted the ambiguity surrounding politics and cultural
change in the colonial era rather than fully endorse Vansina's argument
that the idioms and vocabulary of the precolonial period were largely lost
or were left powerless.[11]

The case of Garner is particularly important to consider in this discussion, because he witnessed the very moment when southern Gabonese
political systems constructed in the Atlantic slave trade era were being
slowly absorbed by the French colonial government. First, it is striking
how intermittently French officials and military power appear in Garner's
accounts, which have never before been examined by any historian of colonial central Africa. Although the violence of French authorities is clear
enough from Garner's writings, the occasional appearance of gunboats and
armed troops collecting taxes did not entirely eliminate the ability of local southern Gabonese political leaders to assert their rights over foreigners or exercise their authority backed by recourse to supernatural forces.
Garner's long residency alongside Gabonese people far away from administrative posts meant he had to negotiate with local political leaders every
day, rather than follow the French government's practice of using sporadic
displays of force as a means of demonstrating control over Africans.

This perspective does not deny the desolation brought by early colonial
rule in Gabon but rather captures how local Gabonese leaders adapted to
these changes by continuing to seek out access to new sources of esoteric
knowledge and by using older frameworks of landlord-stranger obliga-

tions. The ability of French officials to appoint chiefs, collect taxes, and limit trading opportunities did not leave local big men powerless. Such findings back Frederick Cooper's assertion that colonial states needed to rely on violence because their administrations lacked the resources to instill routines of political order.[12] The survival and adaptation of landlord-stranger ties and supernatural connections to political power described in Garner's account also suggests that the links between occult power and the postcolonial state in equatorial Africa described by Peter Geschiere and Joseph Tonda, among others, can be traced back through the colonial period. Florence Bernault has investigated the changing vocabulary of occult forces and their relationship to the state between World War I and the 1950s, and Garner's writings confirm her recognition of the complicated blending of European and African understandings of supernatural forces in the last four decades of French rule.[13]

This chapter looks at several of Garner's essays that involved Western technology, supernatural beliefs, and political disputes. Garner's exhibition of his phonograph reveals southern Gabonese understandings of gender, spirituality, and landlord-stranger relationships. While Garner viewed Gabonese beliefs as proof of African backwardness, these stories permit one to partially reconstruct how local political leaders sought to control foreign objects believed to contain supernatural power. One narrative discusses a long dispute between local landlords with Garner over rent that left the American stranded and shunned by Nkomi people for several months in early 1906. Landlords demanded that Garner pay European currency in monthly installments so that they could pay their poll taxes in cash, rather than continue Garner's previous practice of occasionally offering trade merchandise as gifts. The American, unable to reach French officials, then invented his own supernatural ordeal to frighten his opponents into abandoning the boycott against him. This encounter denotes how Garner and his rivals used a contested vocabulary of political and spiritual power in their struggles with each other without any direct involvement by the French government, and how French impositions such as poll taxes had altered previous conceptions of landlord-stranger relationships. The last section considers how Garner participated in public religious ceremonies as a means of showing respect toward his hosts. These accounts demonstrate how Nkomi leaders presented themselves as the guardians of their American guest and how they in turn expected Garner to look after their interests. The rituals served as a means of showing the value and ne-cessity of local knowledge and political leadership.

Garner's Magical Phonograph

Garner penned an essay, "A Phonograph among the Savages," about his demonstrations of his sound-recording equipment to Fang, Gisir, and Nkomi communities around 1906. Garner presented his tour as an ex-ample of credulous Africans: "Poor, superstitious beggars! I sympathize with them and cannot blame them for their childish fears. To spring at a single bound from the depths of barbaric simplicity to the apex of modern civilization without feeling the shock is too much to expect of human na-ture."[14] Garner chose to enter this crowded field for both ostensibly scien-tific purposes and his own amusement. Capturing the songs and speeches of Africans went hand in hand with Garner's interest in recording the sounds of monkeys. He regularly brought with him music from the United States, as well as blank cylinders that he employed to record local people, although unfortunately none of his Gabonese recordings have survived.

What Garner did not realize was how his tour also reflected long-standing conflicts over the movement of esoteric knowledge and foreign-ers in southern Gabon. By 1906 the old Nkomi monarchy splintered into small districts, in which some Nkomi leaders gravitated toward French administrators while others isolated themselves from the French presence. Decentralized Fang and Gisir clans sometimes accepted French authority but often ignored government mandates or sought to evade troops sent to collect taxes. The passage of goods and valuable patrons through southern Gabon became yet another point of contention in these negotiations.

Prior to the 1880s Nkomi leaders had effectively controlled the move-ment of Europeans and unusual goods in their domains. Indeed, it would be a unique trade item (such as Garner's apparatus) that brought about the end of the Nkomi monarchy. In 1881 a Scottish trader chose to give a luxurious armchair to a rebellious Nkomi clan leader rather than to King Oyembo Onanga, and the king's decision to have the Scot imprisoned for his audacity led to a French military intervention followed by the king's surrender of his sovereignty.[15] Individuals in central Africa seeking to develop prestige and authority sought, as Jane Guyer has argued, to engage in "multiple self-realization," which required them to show their value through unique combinations of goods and knowledge.[16] With the expansion of European commercial and political networks into Fernan Vaz, one can see Oyembo Onanga's anger as an example of what Arjun Appadurai has termed a "tournament of value," in which individuals and groups struggle to impose, reject, or modify radically different understandings of economic value.[17] With royal and clan authorities no longer able to guide the movement and distribution of imported goods and foreign patrons, individuals strove to develop access to foreign knowledge and merchandise to better negotiate this chaotic period of economic and political fluctuations.

By playing and recording Gabonese voices, Garner presented himself as a *nganga*, a diviner who could access the spirit world through the arcane knowledge and unusual power objects at his disposal. Like the mid-nineteenth century explorer Du Chaillu, he promoted himself as an "enchanted white man" who commanded wealth and mystical forces.[18] Yet Garner's introduction of the new and mysterious device was hardly a novel event in Gabon. In the late nineteenth and early twentieth centuries, Mpongwe clans in Libreville regularly employed talismans that came from southern Gabon through slaves and trade connections.[19] This traffic in talismans existed along the central African coast for centuries, although the dynamics of this commerce remain obscure.[20] Beliefs regarding success in trade, arcane knowledge unavailable to local communities, and human sacrifice came together in highly prized power objects.[21] Unique power objects such as Oyembo Onanga's English-made bell in his capital constituted crucial elements to the spiritual power of political authorities.[22] Trade routes had long conveyed new innovations in spirituality in Gabon, whether by missionaries, African Christians, or indigenous religious practitioners. Perhaps the most well-known example of this process was the passage of inland southern Gabonese *bwiti* ceremonies to Fang and Omyènè clans.[23] In 1897 a Nkomi man named Kadinga introduced a new spirit cult known as *odandanga*, into which he had been initiated in the colonial capital of Libreville. When he died in a ceremony designed to initiate others into its secrets, the fifteen Nkomi people who had joined the movement demanded to be reimbursed the fees that they had paid Kadinga.[24]

Garner's machine unsettled southern Gabonese people as much as the failed odandanga movement. In the Nkomi town of Aziambe, the American visitor recorded a speech given to welcome his arrival by the local chief, Olundu. This chief was "widely known in several of the adjacent tribes and a man of great prestige among them, where he carried on an extensive trade in the interest of white men on the coast." After the recording, Olundu came down with a fever that took his life the next day, after Garner had already left the town to go inland. Garner learned of this event after he arrived at another town roughly a two-day canoe ride away from Olundu's settlement. He also observed that the news had spread in villages over a week's time away from Aziambe. Olundu's relatives in other towns apparently viewed the phonograph as *mbuiri* and refused to be recorded.[25]

Anxiety preceded Garner's arrival in other settlements. When Garner tried to convince inhabitants of a Fang clan settlement to sing, he played for them a section of his Olundu recording. The villagers knew of the Nkomi merchant and his untimely demise. While one man cried in Fang, "*Olundu, Olundu a jui*" (Olundu, Olundu is dead), the inhabitants scattered. Garner wrote, the Fang people "insisted that no man's voice could live after the man himself was dead. . . . They demanded that I should take the 'box,' as they called it, and leave the town at once."[26] Some villagers menaced Garner with violence. He managed to persuade the chief to allow him to stay in a house within the settlement and followed his host's stipulation that he leave immediately the next day.

In a Gisir village, local people made comparisons between their own objects imbued with supernatural force and Garner's tool. Most people in the town did not want anything to do with the phonograph once Garner had played back to them their own recorded voices. Garner recorded that three men from a neighboring town "explained they had learned from one of the great 'medicine men' from their tribe that the box contained the head and the heart of a dead man who had become the victim of witchcraft and that it was through the aid of these gruesome accessories that I was able to perform the miracles attributed to me."[27] Southern Gabonese clan ritual specialists, chiefs, and heads of households kept skulls of ancestors in reliquary boxes that were believed to be sources of supernatural power.[28] Rather than being a unique arrival, the phonograph fitted well with local concerns of the spiritual nature of the body as well as the use of reliquaries to hide supernaturally prepared body parts from view. Such an explanation also included a not-so-veiled accusation of sorcery against

Garner. What may have well made the phonograph so threatening was not its novelty, but rather how much it seemed to serve as a bridge between the living and the dead, just as their own reliquaries did.

Not all local people feared the new technology. Some wished to fit it into local concerns about gender, as evidenced in another Fang village. The local chief listened with irritation to a recording of a Gabonese woman singing and laughing. Once the cylinder finished playing, Garner was surprised by what next transpired: The chief "gravely asked, 'Now can you let me hear a woman crying?' I was not only surprised at this singular request but my curiosity was aroused and I inquired why he wanted to hear such a record, to which he calmly replied, 'Because when I hear a woman cry I know that she is conquered and I like to hear that; it makes a man feel good to know that he is their master.'"[29] Women should not laugh before men, the Fang chief added. Given that other villagers also heard this conversation, this may have been a public performance of male dominance, because Fang men before World War I often celebrated their control over women's labor and their ability to obtain wives through warfare and trade.[30] The chief may have used the phonograph as a means of using esoteric knowledge to further bolster his authority over female family members. Other people also wished to use the machine to settle disputes over marriages. For example, people in a Gisir settlement asked Garner to use the phonograph at the grave of a dead man "to see if he could be induced to come back and pay a 'bundle' that he still owed on one of his wives."[31] Gabonese people wanted to harness the foreigner's power object for their own ends.

The varied reception of Garner's machine fits into a broader movement of occult technology and beliefs in the late nineteenth and early twentieth century, before missionaries and administrators could destroy objects. Garner followed trade routes that had long served as conduits of supernatural knowledge and power objects. The ambiguous reception of the new object in many villages testifies both to the respect accorded individuals who guarded arcane knowledge and to concerns about the damage such innovations could potentially cause. These narratives also highlight the attractions and the fears that new, rare occult knowledge might bring to communities facing the anarchic conditions of southern Gabon in the concessionary era.

The Landlords Strike Back:
Struggles over Mobility and Power in Gabon

While many of Garner's tales present their author as a heroic figure always able to outwit his Gabonese counterparts, others expose how dependent he was on his hosts. In 1906 Nkomi families refused to provide him with either food or labor after the death of Garner's patron Anjanga.[32] After noting how "words can't convey an adequate idea of [some Africans'] diabolical cunning," he complained in the beginning of the essay that Gabonese people were adept in "trickery and deceit." What was clear enough were the grim conditions he faced. Stranded on an island in the lagoon, Garner had no weapons and only meager supplies of food. Certainly, he could not doubt the efficacy of Nkomi tactics designed to make their client come to favorable terms.

Garner's troubles commenced with a dispute over the inheritance of Anjanga. He recognized the value of making an agreement to favor a local leader: "Theoretically this is a very equitable alliance . . . [and] relieves the white man of many tedious details in his intercourse with the natives . . . and gives him a definite status among them, gives him prestige and ranks him with the family chiefs, who he can assemble in case of emergency." The death of Anjanga created a feud within the clan chief's family over who would become Garner's next patron. Following matrilineal lines, Anjanga's nephew Ragombo claimed that he was the new guardian of the foreigner. He asked for permission to establish his village on an island where up to this point Garner had lived alone, because according to local practice the descendants of a dead chief had to abandon the deceased man's residence. Yet little was done to move anyone onto Garner's island.

According to the American, a succession struggle had developed between Ragombo, Anjanga's cousin Repero, and Anjanga's mother. This dispute exposes how matrilineality could be an arena of contestation rather than a straightforward means of determining hereditary rights. The mother claimed Anjanga's infant son was indeed the rightful heir, and she would serve as the child's guardian. Repero's claims were based "not on legal grounds, but because of his age and competency." However, Garner declared that the "older and stronger" Repero was "but a born villain, a confirmed liar and an incorrigible brute." Garner held a meeting in which the mother and Repero backed off from their demands, and Ragombo recommenced building a village on Garner's island. Repero eventually convinced his rivals to create a new town on the mainland so that he alone would live near Garner's house.

For the next several months, Repero sought to use landlord-stranger ties to extort a wide range of favors. In some ways, his requests differed little from those of his predecessor. Garner had briefly kept track of Anjanga's requests for tobacco, rum, and hard currency to pay his taxes. Most of these pleas seem to have been interpreted as gifts by the American, but Garner did reimburse Anjanga for supplying him with canoes and paddlers. Anjanga's client did not find the occasional gifts burdensome: "Anjanga is a good sort of native but has little idea of the cost or value of money and has no earthly means of paying it back. He has said several times that he intends on making a canoe for what he owes me."[33] What Garner viewed as ignorance were Anjanga's own efforts to set and evaluate the worth of goods and services and to obtain the scarce currency needed to pay the new imposition of the head tax. Just as royal authorities tried to make visiting traders accept certain prices in the 1890s, Anjanga sought to develop a similar arrangement with Garner.

Repero wanted to place a cash value on the obligations that he felt Garner owed him, and he did so in a way Garner found much more intrusive than the method of his last Nkomi patron. According to Gabonese scholar Pierre-Louis Agondjo-Okawe, all foreigners adopted by a clan had to pay tribute to the *rengondo* as well as to clan leaders, who could end their support if their guest did not show sufficient hospitality.[34] Now that the *rengondo* no longer could act as an arbitrator of price or the rights of residents, aspiring individuals could act in his place. Garner provided some money and goods so that Repero could pay his taxes to the French government. He later discovered Repero had not honored their agreement but had spent the money on rum for his own consumption. Furthermore, Repero refused to reimburse Ragombo. Worse followed: "About the same time I found that [Repero] was working in every covert way to my injury, was creating mutiny among my servants and trying to force me indirectly to pay double and treble the current prices for everything I needed to buy—He was circulating all manner of evil reports about me in order to prevent other natives from bringing anything to sell or doing any kind of labor for me."[35]

Repero's creative redrawing of landlord-stranger relations soon included a monetary provision. When Garner refused Repero's suggestion to pay him a monthly stipend and suggested that Repero leave the island, his landlord declared that in a meeting of Nkomi elders that the assembly had agreed that Anjanga's claims over the island had ended at his death and that the family now wanted a new payment of fifty dollars for the land. Repero added that he would be happy to collect the money or a part of it on behalf of the family and would hold off Garner's creditors if he could not deliver all of the money. Agondjo-Okawe argues that such commoditization of land was a late nineteenth-century phenomenon, although the limited nature of sources does not provide much to evaluate this point.[36] At the same time, it appears that Repero was following precedent in reshaping his agreements with Garner, because clan chiefs could annul previous decisions to allow foreigners to settle on clan lands.[37] Regardless, Anjanga had apparently not made such exorbitant assertions, and Garner struck by threatening to go to the local administrator over Repero's behavior. The multiplicity of legal and political authorities created opportunities for Garner, as well as Repero, to maneuver.

This tournament of value did not end with the American's choice to menace Repero with French intervention. A few days after the Nkomi leader had been rebuffed by his American client, he convinced Garner to lend him a large canoe to take his sick wife to her mother's village a few hours away. The canoe did not return for nine weeks. Meanwhile, not a single Nkomi visited Garner. Infuriated, Garner railed he would send a French gunboat to "burn [Repero's] village to the ground and have him and his whole family taken before the commandant without another hour's delay."[38] Of course, Garner had no more food and no way to actually reach the administrator. Eventually, the American's cook discovered that Repero had set a boycott against Garner over his refusal to pay his debts. Supernatural force would harm anyone who dared break Repero's command.

To overcome these threats, Garner chose to draw on indigenous and foreign methods of enforcing authority. He managed to persuade a passing stranger on a canoe to send word to the local colonial official and announced that he had "white man's mbuiri" that could both detect and kill a thief. He presented Repero with a ginger ale bottle marked with a skull and crossbones and filled with clear water. Then, he idly placed several nails within Repero's grasp on which he had placed a chemical powder that turned red in contact with water. Garner then put some water from the bottle on a cloth and asked each person to put their hands on it. Because Repero had touched the nails, his hands turned red. Garner's rival panicked, and the whole crowd begged Garner for the bottle after he said that Repero would have died if he consciously tried to steal the nails. Garner responded that he would use the medicine to find out who had stolen the canoe. Repero returned the next day with the canoe, claiming it had drifted off. He turned down Garner's offer to hold an ordeal to find the thief, saying the canoe had been found. Gloating over his victory, Garner declared he had successfully taken advantage of Nkomi concerns about sorcery, because "every man and woman of the Nkomi race mortally and morbidly fears every man and woman of the tribe."[39]

Several themes are striking within this account. With the *rengondo*'s authority a relic of the past and the colonial administration a distant yet ominous presence, Repero and Garner had a fair amount of room to negotiate. Still, the colonial government was making its presence felt. Garner linked himself to the coercive power of the state, and one of Repero's ostensive motives was to pay taxes. Garner and Repero shared other tactics as well. They both seem to have made claims of drawing from esoteric knowledge in striving to control the actions of local people. Older tactics by landlords to make agreements with strangers had been adjusted to provide new forms of wealth, but the goal of monopolizing access to a foreigner's goods and services remained. Ironically, it was a Nkomi man who insisted that hosting a foreigner required a payment in money, while Garner wished to rely on reciprocal agreements that did not involve a clear schedule of regular payments. With no *rengondo* to act as a protector of the community, the anarchic situation generated both opportunities and a pervasive sense of fear regarding the uncontrolled use of mystical power. Garner's decision to act out a fantasy of a "white wizard" allowed him to use recourse to occult power as a negotiation tool, no matter how much of a bluff it was on the American's part.

Some years earlier, a Fang village had tried to drag Garner into a dispute with a foreign trading company. Neither Garner nor his erstwhile Fang opponents made recourse to supernatural power, but this incident does indicate again how Gabonese people continued to assert their right to restrict the mobility of foreigners in their domains. Some Fang-speaking men from the Gabon Estuary region, far from Fernan Vaz, lured Garner to travel several hundred miles to their village on the island of Nenguè-Nenguè at some point in the mid-1890s.[40] The offer of a live gorilla, worth several thousand dollars to an American zoo, proved too tempting for the American to ignore. A sign of the growing commodification of the animal trade is that the same tactic had been used by a Fang clan on Nenguè-Nenguè to lure an English trader into an ambush in 1883. The only difference between Garner and his precursor was that a large ivory tusk had served as a lure instead of an ape.[41] Unaware of how Fang clans on the island had robbed and taken hostage European and African traders from the mid-1870s through the early 1890s, Garner set sail on his small schooner.[42]

A group of Fang canoes met Garner's boat, and their armed crews demanded Garner go to their village. Realizing quickly that the gorilla story had just been a ruse, Garner stalled for time. The next day, a tense stand-off took place. Finally, the chief of the Fang warriors told Garner that an African trader working for a European company had set up a store in his town. One week after his arrival, the trader then packed up and left. Thinking Garner was a trader himself, the chief then commanded the American to send another African trader to his store or else he would never be allowed to leave the village. Before Garner's boat was looted, a French gunboat arrived by chance and drove off his foes. "Under the protection of [the gunboat] we got safely to our destination and [we] returned to the coast, where I laid the matter before the French authorities and a little later a gun-boat was sent up and the miserable town wiped from the map of the Congo Français," Garner wrote.[43] Just as with Repero, Garner's anticolonialist sentiments vanished in the face of Gabonese people who affirmed their power over foreigners. However, some of Garner's interactions with Gabonese patrons proved to be less contentious, as the following section reveals.

Incorporating Garner into Nkomi Spirituality

Another event that articulates how Garner and Nkomi people developed a mutually intelligible vocabulary of power and authority is a healing ceremony that Nkomi people held for Garner at a time when he was seriously ill.[44] Unlike the other accounts considered here, Garner did not mock local beliefs but instead praised them as fine examples of hospitality and concern. After a long day of hunting, Garner suddenly felt a series of chills, followed by a crippling fever that left him too weak to get out of bed. Ogoula, the chief of the village where Garner was living, came to discuss the illness and spoke with the visitor in pidgin English. He praised his guest of "three moons" as "the white father 'of our people,'" and then told Garner, "I also know that, in his own land the white man has his own god and makes his own *mbuiri*, and that he does not believe in those of the black man: but when you remain in the black man's country, the black man's *mbuiri* must protect you" (2). Ogoula then asked Garner to undergo a *mpago* healing ceremony to bring back his good health: "To you, perhaps, it may be like the playing of little children, but to us it is not so" (3). The village leader thus asserted both his own responsibilities as a patron and the benefits his local knowledge could have on his favored client. Garner consented to undergo the ritual.

This decision literally marked him as a member of the village community. Ogoula's chief wife, Ipoindu, and all the adults of the village blessed and thanked Garner. The next morning, white kaolin clay lines were placed on his hands and shoes, with the same designs the other inhabitants had on their bodies; this was explained to Garner as a means of showing unity before Anyambie, the creator divinity, and Nyaka, the spirit to be paid homage (5). Nyaka had chosen to attack Garner, and the ritual would appease the spirit by paying it tribute (mpago). Ogoula, his oldest brother, and his son arrived dressed in white cloth, a form of clothing still worn by participants in ceremonies among Omyènè-speaking clans today. The chief began the ceremony by calling out, "Nyaka, Nyaka, why do you bring evil upon my people? What have I done to make my stranger sick? What have my people done that you make their hearts heavy?" (6). This ritual repeatedly expressed the reciprocal agreement between Garner and his hosts, as well as their responsibility to protect him. The villagers and Garner then proceeded to a swampy forested area outside of town, where Nyaka resided. Once at the site, Ogoula placed a series of objects. The group then returned to their settlement, where a series of sacrifices in rum and tobacco were offered to Anyambie in return for his blessing.

In the second set of ceremonies designed to honor Anyambie, Garner was again presented as a member of the community. Ogoula called out as he offered the spirit some rum, tobacco, and palm wine, "In the name of my stranger and my people, sit in my house and partake of all I have" (12). After the end of the rituals, every adult in the town genuflected in front of Garner and said, "*Biambie, ogendo ami*" (Be you well, my stranger) (16). Garner was hardly the first stranger to participate in ceremonies designed to bond foreigners with Nkomi communities: English merchant Trader Horn (A. A. Smith) was initiated into a *mwiri* power association ceremony, and Paul Du Chaillu was eventually named a *makaga*, an official of the *rengondo* assigned to punish criminals and use force to defend Nkomi interests.[45] Just as with his predecessors, Garner was impressed by their gesture: "I had always regarded [their beliefs] as a kin to witchcraft, snake-worship, or human sacrifice [that] inspired my ignorance and fear . . . but I do not think that anyone can fully realize their solemnity or duly appreciate their significance until he or she has been made the special object and beneficiary of them."[46] Although Garner admitted the cultural significance of the rituals eluded him, they did serve to cement the ties that bound him as a client to his Nkomi hosts. He may have relied on technology to pretend to hold supernatural power, but Nkomi people had asserted their command over esoteric knowledge as well, through paying mpago.

Other details from Garner's writings also suggest how his presence corresponded with political and economic turbulence in the early twentieth century. He regularly attended *bwiti* ceremonies held to uncover individuals believed to have used mystical forces to harm others. Mitsogo slaves from the mountains of south-central Gabon held a ceremony designed to cleanse a Nkomi town of evil sorcerers, for example. These ceremonies were sometimes held because of Garner as well, although he never underwent initiation himself. Fierce competition for trade and access to esoteric knowledge among varied Gabonese communities made hosting Garner a potential source of danger. "To-day, for the first time, I learn that the recent fete of buiti in the town was on my account—not to protect me from harm but others who fear harm on my account here exciting the jealousy and envy of other villages and thereby inspiring them to inflick [*sic*] some evil on the people of this village," observed Garner in one essay.[47] Because fears of uncontrolled supernatural forces have so colored Gabonese spirituality later in the twentieth century, it seems likely that mystical attacks to harm rivals with access to foreign traders was one weapon in the arsenal of Gabonese clans and leaders in the nineteenth century. Renoki, a powerful Omyènè-speaking diviner and clan chief on the lower Ogooué River in the 1870s, held rituals to safeguard visitors such as the German trader Oskar Lenz and Africans participating in his expedition; perhaps such rituals also were designed to ward off attacks on Renoki's supporters from competitors jealous of the chief for his European partners.[48]

Garner's negotiations with local political authorities and his esoteric knowledge demonstrate how coastal Gabonese people could inveigle their foreign client into building reciprocal ties. Although the American atheist regularly ridiculed Nkomi beliefs, he ultimately developed a repertoire of performances and a political vocabulary centered on occult force, or *mbuiri*. His understanding of *mbuiri* was somewhat different from those of his African hosts, but his willingness to selectively use this concept to further his own ends resembled the tactics of Ogoula and Repero. Furthermore, Garner was hardly in a position to defend his own rights or impose his beliefs on Gabonese communities with impunity. Repero and Ogoula could mobilize kinship ties and their own assertions of Garner's status as a guest. French administrators and gunboats could not assure Garner's ability to ignore local conventions regarding the rights of indigenous people over strangers. Furthermore, Ogoula and Repero both developed arguments about Garner's position that served their own personal interests. Repero's creative reworking of payments shows how some Atlantic Africans could try to set economic value on land use and services during a time when colonial authorities and rural Gabonese people struggled to impose different forms of assessing economic value on each other. Ogoula's defense of local arcane knowledge to Garner also was a public performance of his authority as a village leader with access to esoteric information unavailable to others. Nkomi people did engage in these negotiations at the turn of the twentieth century, contrary to Gaulme's suggestion that the fall of the *rengondo* monarchy marked the end of Nkomi autonomy.[49] World War I and the timber boom of the 1920s radically altered the playing field in favor of French interests, but even in the concessionary era immediately preceding these changes, Nkomi clan and village leaders still found vocabularies of supernatural force and landlord-stranger relationships to be potent and effective.

Conclusion

Garner's anecdotes about his life in southern Gabon contain information far richer than the paltry and predictable lessons of white supremacy and technological superiority Garner found in them. His narratives of everyday negotiations resemble those constructed by late nineteenth- and early twentieth-century missionary ethnographers such as Robert Nassau and Henri Trilles: the detailed ethnographies these men collected greatly surpassed the limits of their theoretical and methodological frameworks.[50] Unlike these missionaries, Garner could draw from local vocabularies of power and knowledge without making any effort to make Nkomi people alter their own beliefs. Although Repero may have recognized how vulnerable Garner was as a result of his tenuous ties to the colonial government, this anomalous position allowed the researcher access to information that Nkomi interlocutors probably would not have given to priests and administrators. Rather, they may have associated Garner with popular English and American traders who had settled in Fernan Vaz since the 1860s and were remembered by southern Gabonese with fondness in comparison with the rapacious agents of French concessionary companies. Such popularity may account for how willing Nkomi informants were to discuss the conception of *mbuiri* with Garner.

The fluidity and multiplicity of the concessionary era is a central theme in Garner's tales. Systems of determining economic value, the exact nature of the proper relationship between landlords and strangers, and the movement of occult knowledge were all open for contestation in early twentieth-century Gabon. While the *rengondo* authorities were never the absolute authorities of commerce and hidden knowledge in the region, they did serve in the nineteenth century as arbitrators of prices and trade and as intermediaries between spiritual powers and the rest of the community. Although colonial officials and concessionary companies had left monarchy in tatters by 1900, the intermittent and arbitrary interventions of administrators meant that clan and village leaders could act relatively independently. For an isolated and relatively impoverished foreigner like Garner, this anarchic setting meant that he was obliged to use local political vocabularies with far less leverage than missionaries and administrators. His performances of *mbuiri* mastery may have impressed his hosts, but they also made him articulate his demands in ways that placed obligations on himself.

Garner's accounts also highlight the ambivalent responses of southern Gabonese communities to new forms of occult knowledge. Chemical dyes, phonographs, and carbolic acid were not simply evidence of Western technological superiority. They were also examples of the rapid passage of mysterious and potent sources of power throughout much of rural Gabon. Just as *bwiti* rituals and beliefs from southern Gabon were reshaped to fit a wide range of anxieties and concerns among Fang clans in northern and coastal Gabon, so too did Nkomi and other people living near Fernan Vaz put Garner's tools to use for themselves. Local communities could use such mysterious items as a way of launching into discussions of political power, occult force, and the potential results of allowing strangers to stay with them. Missionaries, migrant workers bringing knowledge of *bwiti* spiritual movements, and itinerant African preachers and catechists also spread ideas of arcane spiritual knowledge in twentieth-century Gabon.[51] The ambivalent reactions Garner elicited mirror the varied responses that *bwiti* and Christian followers received from southern Gabonese communities; far from surrendering their own beliefs, southern Gabonese people selectively drew insights and interpretations from this heady mix of practices and objects.

Garner's experiences provide evidence that the concessionary regime and the colonial government did not radically alter local understandings of occult power and political authority in central Africa as dramatically as some have argued. The collapse of formal political structures in Fernan Vaz did not destroy local conceptions of esoteric knowledge and leadership. Instead, local communities and individuals struggled to fit foreign commodities and visitors into indigenous social and cultural networks during a time of tremendous competition and uncertainty. Such strategies continued well after Garner's time. The change lay in the ability of foreign visitors to insulate themselves from Gabonese demands.

Hunter Georges Trial and timber camp owner Marcel Michonnet settled in Fernan Vaz in the 1920s.[52] While their American antecedent fretted over Nkomi landlords and performed a dubious role as a healer and diviner, his French counterparts were fully aware that colonial authorities had now crushed armed resistance and could survey their domains much more effectively than in Garner's day. Trial's hosts begged him to protect them from both elephants and colonial officials who would punish them for hunting big game without permits.[53] Michonnet's family could recruit workers for his timber concession with no worries about open retaliation. Although Garner's deals with Anjanga heralded these later examples of Nkomi clientage and foreign patrons, French residents of Fernan Vaz never discussed threats of boycotts or had to pretend to be masters of mystical force. Instead, they could gain local allies by acting as intermediaries between local people and the terrific and encroaching power of the French colonial state.

While the balance teetered between local authorities and foreigners in Garner's day, they tipped heavily in favor of French citizens from the 1920s onward. Foreign guests no longer felt as compelled to appropriate the vocabulary of *mbuiri* and occult force to obtain what they wished from Nkomi people. Just as Garner's troubles indicated the provisional and intermittent influence of colonial officials and missionaries at the turn of the twentieth century, the relative ease of his successors in holding the upper hand over Nkomi people signaled the encroachment of state authority. Garner rode on the tattered coattails of telegraph wires, isolated administrative posts, and local political vocabulary linked to a collapsed monarchy; his successors operated in a more firmly regimented milieu, in which roads, timber concessions, mission schools, and the urban center of Port-Gentil had helped to undermine the autonomy of southern Gabonese communities. This ambiguous victory of economic change and state authority hardly supplanted indigenous beliefs, as Christopher Gray has noted, but it did make the disadvantages Garner faced in his negotiations a relic of the early colonial past. While Vansina is borne out by the collapse of the *rengondo* position, Garner's papers demonstrate how beliefs and practices associated with secret knowledge and occult power could survive the maelstrom of early colonial occupation.

CHAPTER SEVEN

Aping Civilization

NEW YORKERS HAD A VERY LONG LIST of anxieties to choose from in the fall and winter of 1914. The carnage of trench warfare left hundreds of thousands of corpses scattered on the fields and forests of France, Belgium, Gabon, Poland, and other combat zones. Trade with colonial empires in Africa and Asia had gone into a tailspin. U.S. troops had invaded Veracruz and become involved in the bloody Mexican Revolution. Feminist activist and family planning proponent Margaret Sanger's newspaper, *Woman Rebel*, had caught the attention of many, including the U.S. Post Office, and it was repeatedly suspended. Secretary of the Treasury William McAdoo scrambled to put together a central banking system in the face of panicked European investors desperate to cash in their American investments. The rush to pull funds out of New York banks required the federal government to bail out the city's financial institutions. This chaos compounded the recession that had struck much of the national economy for more than a year. And, in a cage in the Bronx Zoo, Dinah the baby gorilla fought for her life.

Newspapers had made Dinah into a sensation even before she arrived aboard a Dutch steamer in August 1914. William Hornaday, a curator at the Bronx Zoo and a master publicist, had ensured that reporters breathlessly awaited Dinah's appearance. Once safely behind bars, Dinah aroused the curiosity of the public. Richard Lynch Garner had managed a feat few others could boast of before World War I: the successful voyage of a gorilla from African to American shores. The attention that Dinah accrued also reflected on his long and usually futile efforts to claim a place as an internationally recognized expert on apes and Africa.

At the end of the day, it would be the animal's illnesses and antics that newspapers dutifully documented, rather than Garner's achievements. Eugenie Shonnard, a young artist, stayed in the gorilla's cage so that she could properly sculpt Dinah for posterity.[1] The feminist modernist writer Djuna Barnes, then working as a journalist, interviewed the gorilla. When the animal hugged her, she recalled, "I was not embarrassed but only pleased that she had found something in me, as a representative of the women she had come among, to make her trustful."[2] As Dinah became paralyzed from an illness that eventually killed her in early 1915, reporters opined that the comforts of zoo life—which included a maid's outfit and a baby carriage—could not staunch the pain that came with being taken away from her jungle home. Writers argued that the gorilla never could be satisfied with her New York home. Civilization had killed the sad animal.

Dinah's brief American sojourn created opportunities to discuss the relationship between supposedly savage Africa and American society, gender roles, and the daunting debates raised by the application of Darwin's theories of natural selection on human societies. Likewise, Garner's exploits and theories proved fertile material for the popular press from the early 1890s until his death in 1920. Jeannette Eileen Jones has noted how the search in the Progressive Era for the "missing link" between animal and human articulated concerns about race and class in the United States and revealed popular understandings of Social Darwinism.[3] The Bronx Zoo's notorious 1906 exhibition of monkeys in the same cage as Ota Benga, a Congolese pygmy stranded in America, starkly illustrates the ways biological conceptions of white supremacy shaped the display of monkeys and apes. None of Garner's animals shared a cage with the unfortunate Congolese man, despite the commonly repeated assertion that Garner's prize of Dinah was one of the primates locked in the pygmy's cage.[4] Gorillas could also be used to promote the regeneration of white manhood supposedly grown soft from industrialized society, as so aptly noted by Donna Haraway's seminal essay on taxidermist and big game hunter Carl Akeley.[5] American popular culture's enduring fascination for Tarzan speaks volumes on the appropriations of primitive Africa for white male consumption.[6] Other exotic animals, such as the Siamese white elephant exhibited in London in 1884, further led to active debate over the nature of whiteness and empire.[7] These animals were products of transatlantic commercial and scientific ties between continents, but most popular narratives of these animals presented Africa as a foreign, barbaric place.

Narratives involving African monkeys and gorillas could reach well beyond simple expressions of savage Africa separated from white America. Much as with stories of tigers, crocodiles, and other exotic animals, Garner's simians became sites of racial and imperial contact in the hands of the popular press.[8] Constance Clark explores how proponents of popularizing evolution, such as Henry Fairfield Osborn, used images to place natural selection in the American intellectual mainstream and to fight fundamentalist critics.[9] Newspaper and magazine writers appropriated Garner's career and animals for their own ends, which were entirely severed from Garner's own ambitions to become a well-regarded scientist and to denote the similarities between humans and other primates. Instead of showing how Garner harnessed gorillas as a sign of his manly power, writers tended to push Garner out of the spotlight in favor of talking animals. Garner's unusual theories sapped the often ferocious and brutal images of the menacing power of primates and made them better suited for comedy. Popular images and stories of Garner circulated across the Atlantic and even the Pacific.

These refractions of Garner and his animals then served as the means to discuss a bewildering variety of issues that extended well beyond the simplistic narratives of racial conquest and evolutionary hierarchies that came from Akeley's museum displays and Tarzan jungle adventure tales. Different stories on Garner's apes articulated views on immigration, Republican politics, feminists, Filipino prisoners of war, and African American workers in cotton fields. Such variety supports Constance Clark's comment, "Ape-man and caveman humor came from all directions. It could be subversive, but it could also buttress the status quo, seeming to reaffirm traditional cultural values."[10] Descriptions of Garner's career became a collective effort. These stories also implicitly suggested how white Americans could imagine themselves as paternal protectors of animals mistreated by Africans. These transformations radically altered Garner's own hope of attaining national fame. In the early twentieth century, discussions of individual American portrayals of Africans, such as Theodore Roosevelt's safari to East Africa in 1909, have rarely considered the tensions between the individual creators' intent and the popular press.

Narratives of live primates were profoundly shaped by gender, but not in ways that always highlighted brute force. Contra Haraway's thesis, journalists and zookeepers alike presented themselves as paternal protectors of weak, innocent female animals. Akeley's thoroughly dead, stuffed animals were impervious to harm, but live creatures became extremely vulnerable objects of affection. Female gorillas were said to have the same need for a stable family life and male protection as human women. Writers postulated that the inability of female primates to survive for long periods in zoos came from their wistful hopes to return to their own families. Such arguments implicitly reified divides between white American consumers and African locales, yet at the same time readers could identify with zookeepers in imagining themselves to be caring guardians of nature. This message corresponds with the growing emotional ties placed on pets by middle-class families in the late nineteenth and early twentieth century, as well as the growing belief in the importance of conservation.[11] The feminization of primates also shows how women could use foreign commodities to construct domesticated imperial domains in ways not always in keeping with the depictions of African people and fauna that highlighted white male domination.[12] Zoos could empower women as homemakers caring for orphaned animals, just as zookeepers could represent themselves as kindly paternal experts instead of bloody hunters.

These diverse portrayals of animals had to be accommodated by individuals who wished to reach the public. Garner, who struggled to find academic support for his work for decades, recognized that he still could reach a wide audience and attract financial support through the popular press. Editorial impresario Samuel McClure's marketing of Garner's initial trip to Gabon in 1892 and 1893 had clearly left a marked impression on Garner. Although Garner kept trying to correct what he deemed erroneous reports about his life and career, he also knew that newspapers were going to be the best means of spreading his message. Journalistic descriptions of their adventures would shape perceptions of their journeys far more than the authors ever could hope to do. Once articles stopped appearing about Garner, his reputation quickly vanished into obscurity. Without books or academic institutional support, self-proclaimed African experts could easily fade away from collective memory, just as Garner's own efforts to use his Atlantic career to intervene in transatlantic debates over race and natural selection.

Making Monkeys Out of Africans:
Race and Popular Portrayals of Garner

Garner's repeated travels to Gabon furnished a rich vein of material for newspaper writers. Anthropological subjects proved to be a recurring favorite for publications seeking sensational material that embodied negative stereotypes about Africans, and images of the supposed similarities between Africans and apes illustrated popular conceptions of Darwin's theories. While Garner himself firmly believed that Caucasians and Africans belonged to two different species, he never claimed to have found a missing link between humans and other primates in Africa. Yet Garner's rejection of such an intermediary figure between humanity and ape hardly daunted newspaper accounts of his travels. The *Syracuse Evening Herald* informed its readers that Garner had returned from Africa in 1896 and that he supposedly declared, "I am convinced that monkeys talked to each other, and that some of them possess a higher intelligence and a greater fluency of language than many of the African natives." He also had encountered a "gorilla-man" that "would be called in Australia a bushman."[13] Other writers placed Africans and monkeys on the same plane. "Re-volution," a 1893 poem dedicated to Rudyard Kipling and Garner, declared, "One day, in the forest primeval / Where monkeys and pigmies [*sic*] abound / In the drowsy height of the noonday / Arose on the silence a sound."[14]

White readers could use Garner as a means of undercutting differences between animals and humans deemed to be inferior. A woman from Georgia disdainfully attacked the use of vernacular English, the language of "unaspiring, unprogressive, and more or less sequestered communities" by bringing up Garner: "We have had negro dialect and cracker dialect and 'hobo' dialect and 'heathen Chinee' dialect and Chinook dialect, and now Professor Garner is threatening to give us chimpanzee dialect into the bargain!"[15] An 1899 report noted how Garner planned to teach monkeys in Africa to speak using the same techniques applied to the "feebleminded."[16] An anonymous writer at the *Atlanta Journal-Constitution* employed Garner to shove Filipino rebels outside the frontiers of the human race: "If Professor Garner will visit some of our camps in the Philippines he will have no trouble to find monkeys that can talk."[17]

An 1899 story tied together New South economic ambitions and monkeys in ways that further illustrated the brutish nature of African

Americans for readers. A Saint Louis paper announced that Garner "discovered that the monkey could pick cotton as well as the southern negro" and had persuaded a plantation owner in Mississippi to employ fifty trained monkeys in his fields.[18] The planter's results had shown that monkey workers outperformed machines and their human counterparts. "The monkeys are in every way superior to negroes as pickers, and the cost of picking is about one third. I believe this discovery is the greatest that has been made for the cotton planter since Whitney discovered the cotton gin." A thousand monkeys were to be imported from Africa to increase his workforce. New scientific achievements thus could increase economic development and deny the basic humanity of African Americans at the same time. Other newspapers ran the piece as a straight news story.[19]

Garner's apes made appearances outside of the United States, in keeping with the circulation of racist ideas in the early twentieth century in the Anglophone world. British newspapers wrongly reported that Garner had decided to open up a school for apes in Calcutta, where they would be trained to read.[20] Reviewers scoffed at this idea, but they often used the image of talking monkeys to delineate class and racial boundaries. A New Zealand newspaper commented in 1903, "Once get apes to speak [sic], or at least understand human speech and the black and other 'subject' races may find strong competitors in the labour market. With the natural sequence of a 'Monkeys' Anti-Slavery League' [sic]."[21] In the same year, several writers argued with one another about a claim that a chimpanzee said the following Maori greeting to Garner: "*Tena ra kohi, pakeha*" (Good day, stranger).[22] English occultist Aleister Crowley made an anti-Semitic jibe in a reference to Garner. On meeting a Jewish man in New York, he noted, "With all the quickness of his Jewish apprehension, he decided that I was meat for his master, for whom he sent by means of the complicated manual gestures which form the true language of Jews, and, *pace* Professor Garner, of the other anthropoids."[23]

Popular press references to Garner preferred to draw comparisons between primates and racial inferiors over detailed descriptions of Gabonese societies. Garner himself wrote dozens of essays on southern Gabonese cultural and social practices, but hardly any of them saw the light of day. Meanwhile, his articles on chimpanzees and gorillas continued to be published in *The Independent* and other magazines and newspapers, even at the lowest point of his career between 1904 and 1909. It seems reasonable to conclude that editors chose not to accept Garner's more pointedly ethnographic essays. Likewise, Garner's determination to prove that people of African descent were a separate species from other human races never reached the general public, although it is unclear if Garner ever sent his writings out to publishers on this issue.

When Gabonese people entered into press reports regarding Garner, they served as testimony to the American traveler's heroism. Details about Gabonese people are few and far between. Almost all articles referred only to the French Congo, which actually was the title of all French possessions in central Africa, rather than the specific colony of Gabon. Such a generic portrayal of Africa as a dangerous and wild place was part and parcel of most American presentations of the continent in the early twentieth century. When writers did trouble themselves to furnish more details about Gabon, they tended to stay relatively faithful to Garner's own accounts, albeit with some embellishments. As discussed earlier, Garner had boasted about the adulation he received from coastal Gabonese Orungu clans for his hunting skill. Such flattery may also be explained by the fact the French government did not allow Gabonese people to buy repeating firearms, while privileged foreigners like Garner could carry Remington rifles. Some articles contended that Gabonese people had proclaimed Garner "God of the gun," a garbled version of the Omyènè praise title *oga njali* (king of the gun).[24]

Perhaps the best illustration of how Garner could be conceived as a commanding figure appeared in the *Baltimore Sun* in 1909. Articles about Garner appeared somewhat frequently in the *Sun's* pages, largely because his son Harry Garner worked in Baltimore at an insurance firm from the turn of the century to 1919. The article outlined the costs and opportunities available for American big game hunters considering a trip to Gabon. One illustration featured Garner lying in a hammock, carried by five Africans. Europeans in Gabon commonly used this mode of transportation, known as a *tipoye* in French colonial parlance. The tipoye displayed Westerners' control over Africans, and the image of docile and faceless Gabonese workers served to show Garner's authority to American audiences as well. The title of the piece, "What It Costs to Hunt Big Game in Africa a la Roosevelt," tied Garner directly to Theodore Roosevelt's safari to Africa, which had received extensive coverage in American press.[25] Other pieces placed Garner in a pantheon of heroic scientific exploration. One article printed in both the *Los Angeles Times* and the *Pittsburgh Post-Gazette* placed Garner alongside polar explorer Robert Peary and a number of European scientists isolated in remote locations.[26]

During the early twentieth century, U.S. press coverage of Garner thus followed typical conventions of savage Africans and heroic white men. Monkeys could be equated with people of color from around the world, such as Filipino prisoners and African American field hands. Garner himself disseminated such ideas in his own writings. So did proponents of evolution in general, as they manipulated evolutionary rhetoric to espouse white dominance on both sides of the Atlantic. However, stories about Garner and his monkey escapades could veer into territory far beyond anything he originally intended. Journalists seeking to write comic material fused jokes about politicians with Garner. Others even decided to use Garner's talking apes as an opportunity to critique social inequality.

Political Monkey Business

An irate writer defending Democratic New York senator David Bennett Hill in a Christian magazine declared in 1892, "As for the chatter of most Republican papers against Senator Hill, only an experienced specialist like Richard L. Garner could it make it intelligible."[27] For more than two decades, writer after writer turned to the same Garner jokes to deride their political opponents. The *Atlanta Journal-Constitution* declared that Garner would be able to preside over a populist rally because he had already learned monkey language.[28] The editorial staff was fond of Garnerian monkey metaphors for politics.[29] A letter to the editor in Pittsburgh asked if it was a coincidence that the mayor of the city disappeared just as Garner arrived with his monkeys to the zoo.[30]

Some writers played with Garner to make political points that did not necessarily denigrate people of African descent. In 1911 a detachment of African American soldiers arrived in San Antonio. Congress representative John Garner demanded that President Wilson remove these soldiers out of fear that they might challenge Jim Crow laws, as black troops had done in Brownsville, Texas, in a nationally publicized violent protest in 1906. The *Independent* magazine declared, "This is not the Garner who lives with apes in Africa, but another Garner who fusses with Africans in Texas." The editorial supported African American soldiers against the representative, particularly because his district did not include San Antonio. President Taft chose to ignore the demand and kept the African American troops in the city. The magazine editors concurred with the president and praised the soldiers for bearing unjust discrimination with "more Christian resentment than soldierly resentment." For once, it was a white man left "wondering how he happened to make a monkey of himself," rather than white writers making people of color into apes.[31]

Perhaps the most elaborate use of Garner's work came in the *New York Times* in 1910, and its polymorphic negotiations of race did not entirely fit with Garner's more bluntly racist applications. Humorist Wallace Irwin, famous for his caricatures of Japanese people written under the name Hashimara Togo, managed to weave together negative stereotypes of Japan, feminists, and immigrants in a December essay in which Garner's talking monkeys took center stage. Irwin's character Togo is mainly known as a popular expression of negative Asian stereotypes. However, Yoshiko Uzawa has noted how Irwin used the comic persona of Togo to offer sharp critiques of middle-class American society, mollified by the intricate wordplay that Togo's supposedly poor command of English afforded Irwin.[32] Titled "Togo Solves the Mystery of Republican Defeat—Prof. Garner's Monkanzee Helps in the Affair, Which Is Complex in the Extreme," Irwin's essay demonstrates how Garner's popular image could be reformed to create a white male audience of readers contrasted against a range of buffoons.[33]

The piece begins with Togo's introduction to Susie, the chimpanzee with whom Garner toured the East Coast in 1910 to demonstrate the ability of primates to communicate with humans. Susie, "clad in fashionable trousers of suffragette appearance," smoked cigarettes and so loved Republican defenses of high tariffs that she "will clabber with monkey banzais to the top of the chandelier and wave an American flag for 15 minutes." Irwin added that she could not make a clear distinction between a machine Republican or Democratic politician, "which is the correct reply." Garner and Susie then head for a secret meeting with Senator Elihu Smoot and some other notable Republican politicians, where Garner suggests that the GOP import large numbers of monkeys into the United States. If monkeys could speak English, then they should vote, although Susie herself could not take part as a mere woman. The Republicans then interrogate Susie on basic historical facts and discover that socialist activist Eugene Debs had freed the slaves and that "Columbus discovered [America], but John D. Rockefeller was the first to see its possibilities." Garner convinces the Republicans to import one million monkeys to ensure the passage of tariff legislation, but Togo and his cousin Nogi try to stop the plot to fulfill their duty "as patriotic Japan-Americans." Once they sink the ship, monkeys "speaking 12 different European languages" scramble for safety, and one gorilla screams in rage at the Japanese saboteurs in "Tammany language." One illustration accompanying the piece bears the caption "The Average Voter" and features a white man staring at a monkey-faced man dressed in a suit.

Irwin's twists on Garner's story allowed him to play with racial stereotypes and American political debates. Instead of using the image of African monkeys to disavow African American humanity as southern Democrats so often did, Irwin made fun of northern Republicans. Real African Americans in many states were subjected to a battery of tests on historical facts and the Constitution to ensure they could not vote, instead of the laughing matter of Togo's piece. In keeping with other cracks Togo made about the mutual resemblance of Irish immigrants and African Americans, one cartoon gorilla dressed in stereotypical Irish attire waves a club and yells in "Tammany English," a reference to the New York City Democratic machine's reliance on Irish votes. African Americans and Irish immigrants thus could both be seen as lower forms of humanity.[34] Susie's praise for Eugene Debs and John Rockefeller presents an ambiguous political statement; Garner becomes just another comic figure in a theater of conflicting meanings. Likewise, Susie's witty repartee might seem like an implicit endorsement of suffragettes, but the image of Susie is a ludicrous chimpanzee in a large hat, smoking a cigarette.

Another imaginative reshaping of Garner's research with a far clearer set of political morals came with Herbert Casson (1869–1951). What makes Casson's references to Garner particularly impressive is how he cited Garner to support ideologically opposed ends over the course of his career. A Canadian Protestant clergyman and a proponent of Christian socialism at the turn of the twentieth century, Casson made a dramatic about-face from harshly critiquing capitalism to penning fawning descriptions of farming business giant Cyrus McCormick. His later work extolled wealthy entrepreneurs unfettered by regulation, as evidenced by such titles as *Creative Thinkers: The Efficient Few Who Cause Progress and Prosperity*. Garner and his monkeys first appeared in Casson's imagination in an 1899 essay, "What Fools These Mortals Be." Casson praises monkeys, who do not "buy coconuts from the great-great-grandchildren of the gorilla who invented a way to crack them" or "appoint a few monkeys to govern them and then allow those appointed monkeys to rob the tribe and mismanage all its affairs." Richard Garner then is invited to take his talking apes on a tour of "a civilized nation," where millionaires occupy mansions with forty rooms while forty poor workers are squeezed into overcrowded tenement apartments. Once Garner tells the perplexed simian that the masses were content to "endure such evils" and grew angry about change, the chimpanzee responds, "Take me back to the forest, and may the Good Spirit deliver us from civilization."[35]

Such concerns about the unjust nature of industrialized societies vanished in a later reference to the Virginian scientist. "Professor Garner—the man who spent his whole life studying monkeys—once told me that monkeys would gather around a fire in a jungle. . . . But never, by any chance, would any monkey ever put a stick on," Casson noted in a 1926 self-help book for salesmen. The moral was simple—the vast majority of people "work because they must, and they do as little as they dare. They required trainers and supervisors."[36] Garner's lessons now served the Gospel of Wealth rather than the Social Gospel, and the huddled masses that so appalled Casson's monkey of 1899 now were presented as unimaginative boors. Simians could transcend the great divide between nineteenth-century liberalism and the calls for state action to alleviate economic inequalities.

What Casson sought to illustrate for opposed political ends was a common feeling in the Progressive Era: disappointment with civilization, however one defined the word. Famed cynic Ambrose Bierce, one of the most embittered observers of the American fin de siècle, also reflected on Garner's talking animals to critique contemporary society. Noting how Garner wanted to raise animals to the level of civilization, Bierce then took aim at the grim realities of American life. In the name of civilization, whites had employed "the shotgun, the bloodhound, and the faggot-and-stake" and "hunted the Red Indian from cover to cover." Monkeys could learn to enjoy the abolishment "of rest by means of lights and rattling vehicles; of generating sewer-gas and conducting it into dwellings; of loading the atmosphere with beautiful brown smoke." It would be hard to explain to these animals the value of women trapped in tight corsets and by men. Less difficult to understand would be granting them the right to vote so that they could follow "the dictates of the bosses." Bierce reaffirmed his pessimism at the end: "If we fail to persuade the monkeys forward along the line of progress to our advanced position it will be pleasant to have from them an occasional word of cheer and welcome as we are led back to theirs."[37]

Others concurred with Casson's earlier conclusion that monkeys represented a simpler and more ethical society, uncorrupted by the results of modern life. "The more experience one has with the wisdom of men the greater becomes the importance of the knowledge of monkeys. Man talk is getting too complicated, and has too many 'if's' in it," suggested one writer in *Life* magazine in 1901.[38] A decade later, *Life* revealed an excerpt of a conversation between two monkeys about Garner. One monkey warned that younger simians had foolishly started to give away the secrets of their language to Garner, and the results would be disastrous. Monkeys would go to school and then be sold books in their own tongue. Furthermore, the monkey said, "captains of industry [would] put up factories and build railroads and give us employment, so that we could work and earn money to buy the books. I tell you, friend, by that time we wouldn't be any better off than humans."[39] Fictional chimpanzees and gorillas could thus embody concerns about the negative consequences of American life. So did real primates in zoos.

Hairy Ladies and Motherly Guardians: Discussions of Garner's Primates in American Zoos

Garner's longtime opponents in *Truth* magazine first treated his claims as the stuff of jokes. In 1893 a short piece suggested how outlandish his ideas might be: "Syms—Now that Professor Garner has made monkeys talk, what practical use can he put them to? Smyles—He might utilize them at next season's afternoon teas."[40] Gabonese chimpanzees finally did enjoy an afternoon tea, albeit seventeen years later. The New York Zoological Society held a fundraiser on 15 May 1911 where well-heeled donors could enjoy their tea with Susie, the chimpanzee Garner had declared could understand human language. Six other apes "manipulated forks with skill, while Susie, the young chimpanzee recently purchased from the monkey expert, Richard L. Garner . . . [was] the only one of the animals dressed for the occasion. Susie was wearing a new style harem skirt. She ate like a lady." Visitors could also enjoy the unveiling of a white rhinoceros head killed by no less an august figure than former president Theodore Roosevelt.[41]

Why did such exhibits draw so much attention? Donna Haraway has shown how naturalists like Carl Akeley hoped to renew white manliness by exposing zoo and museum visitors to powerful gorillas untouched by the degenerative effects of modernity.[42] Indeed, some commentary on Dinah supports Haraway's contentions. A children's magazine proclaimed at Dinah's death, "What red-blooded boy has not thrilled at the sound of that mighty [gorilla] chest, beaten upon by hairy paws, and that voice that made the lion's roar seem like a treble pipe!" But, the author had to admit that the animal's calm behavior hardly fit common descriptions of gorillas as fierce monsters.[43] The sight of Dinah in a fur coat left one writer a bit unimpressed: "While it couldn't be said she looked pale, she did look puny."[44] Popular acclaim of Dinah, Susie, and their fellow primate Ningo did not fit the profile of brute strength and unvarnished wildness that Akeley carefully constructed. Live animals faltered from illness rapidly, and it is hard to imagine that a gorilla in a maid's outfit in a baby carriage could have mustered up the call of the wild for any visitor. Instead, zoo-keepers could play out their own idea of a civilizing mission. Apes could be dressed as humans and be taught to eat with knives and forks, just as missionaries happily displayed postcards of African converts dressed in suits and modest long skirts.

Zoo depictions of Garner's female gorillas constituted an ambiguous victory of domesticity and gender conventions over savage nature. Certainly, David Hancocks has a good point in noting how chimpanzees displayed in ridiculous outfits in early twentieth-century England constituted "a source of relief for many visitors, especially in an era that exhorted racial differences and in societies that believed that Whites were superior to people with darker skins."[45] Instead of highlighting savage violence, zookeepers and reporters represented their charges as frail, helpless children and women. Mademoiselle Ningo, the first gorilla Garner brought to America on behalf of the New York Zoological Society, almost immediately became ill once she entered the Bronx Zoo. Keeper Raymond

Ditmars offered her milk as well as an alcoholic cocktail to try to raise her spirits. Three chefs prepared meals for the "dainty maiden," to no avail.[46]

Ningo lived only ten days in New York before she passed away, so she did not have the chance to wear women's clothing or ride in a baby carriage as did Dinah during the gorilla's yearlong stay at the zoo from August 1914 until her death. Dinah "daintily ate ice cream and cake and drank milk from a glass. At meals she invariably wore a napkin and used a knife, fork, and spoon."[47] Once a bone disease left Dinah partially paralyzed, Ditmars had her walked daily in a baby carriage with a fox terrier to keep her company.[48] These ideas came from assistant zookeeper Fred Engelholm, who treated Dinah like a sick human child. "'Does oo want to see dap itty buffalo?' asked Nurse Engelholm yesterday. 'Hang it,' he continued, 'I thought I was talking to a kid,'" noted a *New York Times* article. The same piece noted how Garner's kind treatment of Dinah had made her "a spoilt infant." Engelholm provided an entertaining spectacle by reversing gendered conventions as a feminine protector.[49] After Dinah died, one article observed, "Dinah was scarcely a captive. She was more of a pet, almost a child, to the keepers."[50] Such presentations again bear witness to Nigel Rothfels's remark that Hornaday and the Bronx Zoo "underscored a narrative of the civilizing and urban bourgeois ambitions for the world—through the educated and philanthropic activities of the best members of human society, the rest of the world's people and all the world's animals could find shelter in benevolent care."[51]

Most articles preferred to find lessons that affirmed male-headed families and female vulnerability. Newspaper articles blamed the high mortality rates of the Bronx Zoo's gorillas on "nostalgia," but their particular interpretation of yearning resembled a longing for an idealized version of family life menaced by industrialized society. Neither Ningo nor Dinah ate most of the foreign foods they received from zookeepers. William Hornaday later blamed such stubbornness on the low intelligence of the particular gorillas he received before the 1920s: "Personally [Dinah] was a stupid little thing. . . . Her more pronounced and exasperating stupidities were shown in her refusal to eat, or to taste, strange food, even when very hungry."[52]

Most press reports pointed to the sadness of animals separated from their kin as the root cause. A wire service article published after Dinah's death noted how the primate received every luxury but "apparently, she never forgot her early life, for which all the luxuries of civilized existence offered no adequate compensation." The article went on to describe how Dinah's father jealously guarded his family at night as they slept in a tree. Two images accompanied the story: a photograph of Garner holding Dinah, dressed in a girl's outfit, and a crude picture of an adult gorilla with the caption, "Little Dinah remembered her tree nest and herself in Mamma Gorella's [sic] hairy arms. Papa Gorilla sat at the tree base and waited for lions and other things while big brother did acrobatic stunts on the branches. Dinah was a happy baby then."[53] Such a reproduction of domesticity furnished so much love that Dinah could not survive without it. Just before Ningo's death, the *Washington Post* opined, "[Ningo] will not eat, and there is a faraway look in her eyes as she longs for darkest Africa. . . . Sixty of her brothers and sisters have left Africa and started for New York, but homesickness and seasickness killed them all."[54] Too much literal and figurative distance existed between Africa and America to allow animals to survive. The Atlantic Ocean now was a chasm between the United States and Africa, not the means by which the two regions came together.

At least one writer argued that the essential feminine traits of sentimental bonds explained the failure of female gorillas to stay alive, while others asserted how male zookeepers could indeed be motherly. Mary Corthope wrote an essay about Ningo in the *San Antonio Light*, titled "Nostalgia, and Its Nepenthe." She contended that Ningo's longing for her origins belied how "homesickness is almost as unfashionable to men as appendicitis is fashionable for women. It has been evoluted [sic] from the male sex." Corthope mocked men who "prate of home and are anxious to tell strangers how much they love it," but who rarely act on it. Women, whether gorillas or Napoleon's own Josephine, were "home-living by instinct," while men had "appropriated 'wanderlust' as a word applicable to their own sex." Such innate differences could never be crossed, even though the sufferings lay squarely on females of the species.[55] However, a column titled "Daddy's Bedtime Story"—itself a claim to a caring kind of manhood—recounted to readers how a zookeeper, "a great big man, who loves animals very much, and [who] was very, very sorry for poor sick Mistress Dinah," gave the gorilla daily rides in a baby carriage.[56] Garner's gorillas could thus serve as evidence for inherent differences between the sexes or as proof that men could be kindhearted caretakers.

Several authors likened rebellious simians to feminists rather than fe-males desperately in need of family support. Once Ningo rejected her milk and brandy cocktail, a reporter took aim at female supporters of temper-ance. "The prohibition party gained a recruit yesterday," began the brief article.[57] The *Fort Wayne Journal Gazette* found Dinah's unwillingness to eat all too similar to another brand of rebel. An essay on forced feeding at the Bronx Zoo described Dinah as a "'no-terms' striker. Having tired of life, apparently, she went on a hunger strike as the shortest way out." The article then mocked her stupidity for denying all the "succulent delicacies obtainable." Feminist activists had gone on numerous hunger strikes in the early twentieth century, and thus the piece compared the supposed ir-rationality of such methods with Dinah's foolishness.[58] Feminists rejected male authority, like unruly animals that refused the wise tutelage of the zoo staff. This comparison also obscured the use of violence to keep ani-mals in line at the zoo—yet another example of the mystification of the slaughter and abduction of exotic animals in the trade that kept the zoo in business.[59]

A small number of female observers subtly promoted alternate presenta-tions of Dinah that promoted the gorilla as an independent woman in her own right, who evaded the easy explanations of her behavior provided by Garner and the Bronx Zoo staff. Although the sculptor Eugenie Shonnard (1886–1978) became relatively well known for her work dealing with south-western Native American communities, the New York Zoological soci-ety commissioned her to make a bust of Dinah. For Shonnard, a woman trained by legendary French artist Auguste Rodin, such a task was not something she had in mind.

However, Shonnard agreed to go into Dinah's cage. There, she devel-oped a strong bond with the gorilla. In a 1964 interview, she remembered, "I suppose I sang to [Dinah] or hummed, or whistled a little bit. She liked it, and she demanded it. So everyday when I would go she would demand that I would sing to her before I could handle her. She would stand on her hind legs, stand up straight. She was about like this and thump, lick, umphhh . . . that means come on and sing. And we did, before we did any work." Although Dinah tried to wreck Shonnard's clay, the artist regretted the end of her work. A short time after the bust was completed, she visited Dinah. Shonnard told the interviewer,

Well, she was out there in the garden in a big enclosure and I saw her. I didn't have time to do anything but to wave to her, "Hello Dinah," as I ran over to see Dr. Hornaday. Then I had to run to catch my train. So I merely waved to her. Well, that's that. So I said to Uncle Cliff when I got home, "Uncle Cliff and Beatrice," I said, "Now I'm going down to spend the whole afternoon with Dinah." It was about a month later. She never allowed me to see her face again. She was so hurt. She put her arms around her face."[60]

This kind of emotional vulnerability is a sharp contrast from Garner's own accounts of individual primates, in which animals accept his authority as a wise patriarch.

Shonnard wasn't the only one who considered herself to have a special relationship with the animal that crossed the boundaries of language and species: so did Djuna Barnes. Before Barnes's literary works and friendship with James Joyce and other high modernists secured her reputation as a writer in the 1920s, Barnes struggled as a reporter. Her visit to Dinah's cage in the company of Garner was a typical publicity stunt for all concerned and demonstrated how Barnes reshaped such stunts as a foundation for individual emancipation.[61] Initially apprehensive, Barnes decided Dinah was an enigma no human could unravel. "I found," she wrote, "for I had come to study her, that the largest and most splendidly satisfying thing in Dinah's life is herself. She would rather stand well in her own estimation than upon a social footing." Such a statement undermined both Garner's efforts to use Dinah to promote his theories and the more conventional gendered personas attributed to the gorilla. When Barnes questioned Dinah about her views on the United States, she laughed wildly. She also noted Garner's inability to sum up Dinah's personality and concluded, "It had just been borne in upon [Garner] that even here Kipling's remark about the female of the species holds true."[62]

Such an individualistic take on Dinah's femininity challenged typical views of Garner's primates as weak and defenseless members surrounded by men with the expertise to protect them and translate for them. Barnes also used Dinah to critique New York life by announcing that the gorilla had said, "I was quite grieved to observe that the sun in New York has no chance, and the moon is only a past memory. I couldn't make out whether it was daylight or electricity."[63] Barnes thus had remade Dinah into a proud individual unbound by social graces, much as she viewed herself.[64] Like in Shonnard's account, Barnes undercut Garner and the Bronx Zoo's efforts to define how gorillas belonged to fictive families overseen by men. Just as Barnes's self-presentation suggested her individuality, so too did Dinah's behavior, which could not be entirely controlled by her male handlers.

Taken in its entirety, press coverage on Gabonese primates belonged to a much broader trend in the gendered nature of animal-human relationships in the early twentieth century. The rising importance of emotional attachments between animal pets and human owners had made their way into zoos. In the nineteenth century, discourses of feminine weakness and sentimentality could be placed on even the gorilla, commonly presented as a monstrous example of wild African nature. Milk, carriages, and gentle caretakers could tame even the most savage of animals. Such anthropomorphism justified the display of these primates, even though many viewers contended that the basic differences between Africa and North America precluded their survival. Furthermore, representations of female gorillas linked conservation efforts to broader projects of spreading ideals of American domesticity abroad, fostered by zoos staffed by trained experts. Such an entanglement foreshadowed how zoos became spaces where stable, male-headed families held sway over innocent animals, as projected in Disney films, the television show *Daktari*, or more recent programs such as *Animal Planet*.

Garner himself had little control over these representations. Yet again, his own individual aims of presenting himself as a masterful father of primate children rapidly dissolved once it reached newsprint. Gendered spectacles of endangered femininity and motherly zookeepers pushed Garner's own narratives of heroism and detached scientific inquiry by the wayside. Even his theories of animal language did not become visible. Neither did the background of the primates nor the Gabonese people who actually supplied them to Garner. Generic visions of darkest Africa swallowed up the actual details of the business transactions that put animals in Garner's hands, and then anxieties about femininity in the United States took center stage as Garner's own ambitions to make these animals the cornerstone of his own celebrity vanished.

Conclusion

Scholars have identified how representations of primates in the Progressive Era were intimately bound to popular evolutionary theories as well as to constructions of empire and race. Even so, the anarchic deployment of Garner and his talking monkeys exposes how widely varied concerns and meanings placed on simians were in the early twentieth century. The Virginian researcher relied on the popular press as a means of outmaneuvering his institutionally based scientific critics. But, the incongruous idea of talking monkeys made Garner's animals harmless, comic figures instead of fearsome creatures, which did Garner few favors as he presented himself as a heroic warrior for science in the Gabonese rainforest. He and his animals became roaming signifiers, easily placed into new configurations that blurred the line between fiction and nonfiction.

Racial and class hierarchies colored popular depictions of Garner's talking monkeys, even though they did not always match common comparisons between black people and apes. Monkey talk certainly could be used to separate white Americans and Anglo-Saxons from racial and cultural others. However, Wallace Irwin and some other writers used links between simians and people of color as a starting point for more complex and ambiguous ruminations on race and politics, where Irish voters might stand in for African monkeys. Still other observers roped monkeys into fights against pompous politicians, capitalist excess, and even racial discrimination. Satirical deployments of Garner's animals could move far beyond crude stereotypes, although their comic elements played off the biological understandings of white supremacy so prevalent in Progressive Era America. Also, the cultural and political contexts of Gabon simply vanished against the backdrop of darkest Africa, as did the Atlantic commercial networks that turned primates into just another commodity to be extracted from African environments. No matter how imaginative popular versions of Garner and his animals might be, they allowed no room for Gabonese perspectives and knowledge.

Gender concerns set the contours of the discussion of Garner's animals. Officials could be made into mere monkeys, thus demonstrating their failure to live up to standards of honest manhood and sober living. Sickly and weak female gorillas brought by Garner to the Bronx Zoo prompted debate over the nature of femininity. Zoo curators and keepers promoted a fantasy of loving animal children and watchful guardians, but the quickly fading fortunes of Garner's gorillas placed their representations in question. High mortality might represent the vulnerable nature of women or the inherent inability for Africans to survive (white) America. Either case excused the zoo from accepting blame for the deaths of the animals. While some popular writers enjoyed the gender-bending spectacle of hulking nurses pushing gorillas in baby carriages, Djuna Barnes and Eugenie Shonnard poked holes in the zoo's official presentation of stubborn and dumb gorillas unwilling to obey their male owners. In these dramas of femininity, Garner's own self-portrayal as a surrogate father to his animals again was pushed aside. Garner may have prepared his animals for their transformation at the hands of zookeepers by stripping them of their Gabonese contexts, but reporters extricated animals from Garner's own descriptions. In the end, Garner remained a mere supplier of sensational material for others to manipulate for their own purposes, and in doing so they articulated the varied political, racial, and gendered divisions of American society in the early twentieth century.

CONCLUSION

RICHARD GARNER'S LEGEND SWIFTLY FADED after his death in 1920. Even with the wave of popular images of apes and evolution that flooded through U.S. newspapers with the Scopes trial of 1925, the unconventional researcher's career as a primate dealer and evolutionary theorist drew little attention. Academic researchers such as Ada and Robert Yerkes briefly mentioned Garner in their treatises on apes. Usually, university scientists were quick to distance themselves from such an undisciplined and untrustworthy figure, even as they credited Garner as a pioneer in his observations of primates in the field. Yet his many idiosyncrasies and his trailblazing work with apes ensured he never entirely disappeared from view. In the last two decades the outpouring of scholarship on primate behavior and the history of apes in American culture has conjured up Garner again, especially as historians of science have become more and more interested in unconventional scientists.

Garner defies clichés of southern piety and of fundamentalist Christians seeking to ban evolution from high school textbooks. However, the Virginian tried to reconcile his thoroughly peculiar ideas on religion and science with more typical views on race and gender. New South ideals of social advancement through scientific improvement, efforts to convince northern audiences of the value of Jim Crow, and rugged individualism all shaped the contours of Garner's eccentric quest for fame and influence. Electronic gadgets, museum connections, and African travel narratives allowed Garner to combat negative perceptions of his lack of formal education and his humble origins. Garner promoted a white southern version of anti-imperialist racism that conveniently whitewashed his own personal investment in European colonialism. Even if his dependence on the popular press ultimately left Garner with far less control over the reception of his claims than he liked, his case allows for a more nuanced understanding of the idiosyncratic nature of individual American conceptions of Africa in the early twentieth century. While many of his conclusions did not stand out from the current of demeaning stereotypes about Africa and people of color, Garner's unusual agendas indicate how Africa, like the American empire, created spaces for many middle-class Americans to reinvent themselves and express their concerns about race and gender. Whether or not he belonged to an extensive group of southern atheists is hard to determine, but his example does suggest that a scientific community opposed to organized religious belief could take root even in southern Appalachia.

Aware of his vulnerability as an independent scholar with a limited education, Garner followed the demands of evolutionary scientific controversies all the way to central Africa. Garner positioned himself at multiple peripheries of the intellectual, economic, and political currents that linked central Africa to North America. He recognized how debates over animal communication and natural selection linked the United States and Europe, and he made his presence felt in London, New York, and Washington. While many of Garner's critics raised doubts about his true worth as a man, none of them dared to raise themselves out of their armchairs to observe primates in the field. Zoo visitors, museum directors, individual patrons, university scientists, and journalists conversed with one another about evolutionary ideas on both sides of the Atlantic. All of these disparate actors wanted to know more about the links between humans and other primates, and Garner delivered by providing these academic and popular audiences with gorillas and African spectacles. Garner's life exemplified the Atlantic nature of evolutionary ideas, as well as the commercial flows that sent specimens so in demand in North America and Europe.

Gender anxieties and concerns shaped Garner's transatlantic career, just as these issues set the contours of scientific research and Atlantic exchanges of knowledge in other contexts. British and North American readers argued about Garner's masculinity as a means of determining whether or not he was a credible witness, while Garner considered French officials to be failures based on his own standards of white male honor in the U.S. South. It is a particularly great irony that Garner, a man who wanted to use his research to prove the firm separation of races in contrast to Darwin's defense of a single human species, managed to adapt to southern Gabonese gender conventions. His self-representation of a paternal, wealthy patron with access to supposedly supernatural forces made him an impressive figure to his Gabonese partners. By accepting this role, though, Garner became vulnerable to the demands of indigenous communities, who expected him to conform to their expectations of a generous and knowledgeable guest who respected indigenous authority. Meanwhile, American readers expected Garner to provide heroic tales of conquest, and indeed he wished to assert his own heroism in contrast to his academic critics.

Popular and academic audiences sought out primates and adventure tales, and so knowledge and animals traversed the Atlantic from Gabon to Europe and the Americas. Garner brought a late nineteenth-century North American sensibility toward animals that highlighted their loyalty and their intelligence, and he discarded Gabonese ideas regarding gorillas and chimpanzees as threats to fields or as mysterious and potentially hostile forest dwellers. Unlike mid-nineteenth-century travelers like Paul Du Chaillu, who presented primates as examples of savage Africa, Garner's animals were innocent victims in need of defense from barbaric and callous Africans. During the next century, similar arguments would be made time and again by Westerners. Yet Garner's respect for Gabonese informants' knowledge about animals is striking, given his prejudices. Neither his critics nor his supporters in the United States paid much attention to Garner's imperfect presentation of Gabonese perspectives and usually took wide liberties with Garner's narratives of animals to articulate a bewildering range of comments on race, class, and gender. Not only was the specific Gabonese context from whence these animals came from largely lost, but Garner's own ambitions and opinions were often diluted, if not entirely eliminated, in discussions of his work. Primates may have moved along Atlantic commercial routes, but African knowledge about them was largely remade or silenced en route.

Garner did not enjoy the luxury of his Anglo-American readers, who could ignore his Gabonese partners in collecting and researching primates. Like many previous scientific travelers in the early modern period, Garner had to develop creative methods to convince indigenous people to give valuable knowledge. However, he witnessed a radical change in how knowledge was collected in Africa. Prior to the consolidation of French authority after 1899, southern Gabonese people had used access to primates and other animals as bargaining chips in a broad range of commercial agreements. Foreigners had to make concessions to local leaders and enter into reciprocal agreements if they expected indigenous communities to provide them with valuable knowledge about plants and animals. Concessionary companies and the French colonial state irrevocably altered this arrangement. Henceforth, the colonial state's alliance with European firms and scientific institutions would firmly place Gabonese people as junior partners in the exchange of specimens and ideas. Written permits and logistical support from colonial administrators, rather than Garner's willingness to be initiated into local supernatural beliefs and his connections to the fading Atlantic ties between Gabon and the Americas, would now shape how research was conducted in Gabon. Older mutual exchanges of knowledge and material goods that had developed in the slave trade era had come to an end. At the same time, French scholars and researchers became the main scientific authorities on Gabonese fauna, and their discussions almost never included references to the resolutely Anglophone Garner.

As the Virginian disappeared from the forefront of primates and evolution, the social and economic order of southern Gabon unraveled. From the 1920s to the 1950s the consolidation of French colonial authority and the rise of French timber and oil companies supplanted older ties between Gabon and the Americas. While Gabonese people recognized Garner as a partner in long-standing Atlantic networks fashioned in the heyday of the slave trade, Europeans living in the same region just a few years later encountered a very different situation. Georges Trial, a French timber clerk who transformed himself into a self-proclaimed modern primitive and a professional big game hunter in the late 1920s and 1930s, viewed Fernan Vaz as an idealized world of noble savages soon to be dispatched by the banal dullness of European modernity. Nkomi and Gisir community leaders clearly had become the subordinates of French administrators. The kind of autonomy Garner had enjoyed as a semi-independent figure had become a thing of the past. Gabon's international orientation had become firmly fixed on Marseilles and Paris, not Brazil or North America.

Garner's project holds great significance to historical understandings on popular evolution, scientific research in a postslavery Atlantic context, and the relationship between Western interest in primates and colonial Africa. This book has used Garner's idiosyncratic career to follow the shifting

African and North American contexts of primates, from rare commodities in Gabon to tools used by the popular U.S. press to establish a struggling southern scientist's career. Previous studies of collecting and displaying primates, bound to the still looming divisions among historians of science, U.S. cultural history, and specialists of colonial Africa, have not done justice to the interplay between African, European, and North American participants. While recent studies of science in the Atlantic world have brought together the transfer and remaking of knowledge, they have failed to look at the circulation of knowledge from Africa to the Americas and Europe after the end of the slave trade. Too often, U.S. and European historians of science have substituted pious contrition for the role of colonialism and ethnocentric prejudice in shaping the channels of knowledge from Africa to the West, without bothering to explore the varied African contexts in which animals and information were exchanged and circulated. Douglas Candland's 1995 discussion of Garner's research exemplifies this absence. While certainly well intentioned, it laments the lack of African agency in Garner's accounts instead of investigating how Gabonese people viewed, traded, and used Garner and his need for animals.[1] Such neglect is not merely a fault of academics: it contributes to the ignorance that has allowed Americans to indulge in their fantasies with little regard for the needs and aspirations of Africans.

The end of formal colonial rule and the growing environmentalist movements of the late twentieth century led Americans back to Gabon, and the messages disseminated about Africans by Garner still held resonance. A century after Garner lived in Fernan Vaz and pretended to wield supernatural power like any Nkomi *nganga*, American participants in the television program *Survivor* competed with one another on the coast of southern Gabon in teams named after Gabonese ethnic groups. They searched for idols that granted them protection, enjoyed the spectacle of wild elephants, and savored the heroism of African wilderness. Just out of their range of sight, Gabonese troops monitored the isolated setting to ensure that American producers had full freedom to set in play their own fantasies of savage Africa. Garner hardly had a fraction of the budget allotted to an American television production, but he too invented his own kingdom of awe-inspiring animals on Gabonese soil that relied on centralized state domination over indigenous communities. Glossy magazines and colorful websites detail battles of well-meaning Western environmentalists fighting to save animals from the menace of callous Africans in the early twenty-first century, echoing Garner's own idea of conservation as a sign of white superiority. Although Garner's conception of race and animals hardly escaped his piles of typed notes, messages uncomfortably similar to the Virginian writer's proclamations had indeed taken root in American society.

Shortly before Christmas 2008, a plane arrived in the town of Omboué, the district capital of Fernan Vaz that had been Garner's home in 1917 and 1918. One carried a group of foreign tourists immediately met by members of Operation Loango, a nongovernmental organization that holds a concession over much of the nearby Sette Cama and Iguèla Lagoons. Wealthy Gabonese people have found a common understanding in developing ecotourism with Operation Loango and the Wildlife Conservation Society, the latter-day version of Garner's old patrons at the New York Zoological Society. Garner's dream of bringing well-heeled Americans to Fernan Vaz had finally come to pass. The dozen or so tourists were quickly whisked away from the airport on a street lined with rusty tin-roofed shacks. While these pampered visitors could indulge in the idea of wild nature that had so enthralled Garner, their stay was largely insulated from the demands of local people. The postcolonial government's success in quelling local opposition to its control over natural resources and to the movement of foreigners along the lagoon was a testament to a century of centralized state power. Garner witnessed the messy beginnings of this process, in which he still had to wrestle with indigenous understandings of land rights and power to survive.

My research also depended on Gabonese informants and expertise, and I fear far too much of their own understandings of the colonial past have been sacrificed on behalf of my own academic agendas in this book. The name Garner appears to be entirely forgotten in southern Gabon. His writings on the region moldered in North American archives and in yellowing newspapers, and the cultural practices he described have undergone radical changes in the last century. For my informants in the twilight of leader Omar Bongo Ondimba's reign from 1967 to 2009, the history of oil production in Fernan Vaz was the real story that I should ask about, instead of an odd American dead nine decades before my own arrival. Understanding American conceptions of human-animal relations is hardly a topic that most Gabonese living in poverty deem to be important. Many

Gabonese concur with the anger one young man expressed while watching Belgian tourists scramble to see precious pachyderms in July 2007, as he shouted, "Fuck your elephants! I'd kill them all if I could." Writing transnational histories may shed light on unexplored global and Atlantic connections, but they can silence other issues of vital interest at the same time. My own hope is that this work serves to underline the longevity of links between American perspectives of nature and the struggles of Gabonese people to control their destinies. Perhaps this desire is even more unrealistic than anything dreamed up by Richard Garner. Yet a better recognition of how Americans have tried to make Africa the springboard for their own self-understanding is one step in the long struggle to dismantle the inequities that exist between Western countries and African people.

NOTES

Abbreviations

CGE	Collins-Garner Expedition
JC	Journaux de la Communauté
NZP	National Zoological Park
RLGP	Richard Lynch Garner Papers
RP	Rapports Politiques
WHP	William Hornaday Papers

Introduction

1. "Roast on Gridiron: Nation's Leaders Victims of Newspapermen's Quips," *Washington Post,* 19 February 1911, 1; "Gridiron Club at Big Banquet and Make Merry," *Oakland Tribune,* 19 February 1911, 39; Dunn, *Gridiron Nights,* 245–259.

2. Schaffer, "Golden Means"; Parrish, *American Curiosity*; Schiebinger, *Plants and Empire*; Schiebinger and Swan, *Colonial Botany*; Safier, *Measuring the New World.*

3. Numbers and Stenhouse, *Disseminating Darwinism*; Jones and Sharp, *Darwin in Atlantic Cultures.*

4. Desmond and Moore, *Darwin's Sacred Cause.*

5. On the rise of zoos and their links to colonial expansion in the nineteenth century, see Croke, *Modern Ark,* 140–155; Rothfels, *Savages and Beasts*; Hancocks, *Different Nature,* 33–54; Baratay and Hardouin-Fugier, *Zoo,* 113–130.

6. Bridges, *Gathering of Animals,* 346.

7. MacKenzie, *Imperialism in the Natural World*; Fairhead and Leach, *Misreading the African Landscape*; Wylie, *Starving on a Full Stomach*; Zimmerman, "What do you really want?"

8. Shapiro, *Appalachia on Our Mind*; Batteau, *Invention of Appalachia.*

9. On the Scopes trial, see Edward Larson, *Summer for the Gods*; Conklin, *When All the Gods Trembled*; Moran, "Scopes Trial and Southern Fundamentalism"; Clark, *God or Gorilla*.

10. Israel, *Before Scopes*; Lienesch, *In the Beginning*.

11. Clark, *God or Gorilla*, 98.

12. Numbers, *Darwinism Comes to America*, 58–75; Numbers and Stephens, "Darwinism in the American South."

13. Gardinier, "American Presbyterian Mission in Gabon"; Gardinier, "American Board."

14. Cinnamon, "Robert Hamill Nassau."

15. Newby, *Jim Crow's Defense*, 14.

16. Love, *Race over Empire*.

17. Fredrickson, *Black Image*, 229, 263–266; Fitzgerald, "We Have Found a Moses"; Jackson, *Science, Race, and the Case*, 27–35; Fleeger, "Theodore G. Bilbo"; Ward, "Richmond Institution."

18. Kaplan, *Anarchy of Empire*, 110.

19. Stecopoulos, *Reconstructing the World*, 18–38.

20. Jones, *In Search of Brightest Africa*; Sundiata, *Brothers and Strangers*.

21. Duignan and Gann, *United States and Africa*; Hickey and Wylie, *Enchanting Darkness*; Bederman, *Manliness and Civilization*, 207–213; Jacobson, *Barbarian Virtues*, 105–117; Kasson, *Perfect Man*; Mayer, *Artificial Africas*; Keim, *Mistaking Africa*.

22. Redkey, *Black Exodus*; Angell, *Bishop Henry McNeal Turner*; Stein, *World of Marcus Garvey*; Sundiata, *Brothers and Strangers*; Barnes, *Journey of Hope*.

23. Jones, *In Search of Brightest Africa*, 132–176.

24. Mills, *Discourses of Differences*; Blount, *Travel, Gender, and Imperialism*; Early, "Unescorted in Africa"; McEwan, *Gender, Geography, and Empire*; Harper, *Solitary Travelers*; Brisson, "Fish and Fetish."

25. Pratt, *Imperial Eyes*, 209–213.

26. Musselman, "Plant Knowledge at the Cape"; Beinart and McGregor, *Social History and African Environments*; Jacobs, "The Intimate Politics of Ornithology."

27. Safier, "Fruitless Botany," 206.

28. Information on Garner's siblings is drawn from the Garner family Bible. See the entire list of Garner's siblings: Janice Johns, "Samuel Garner Family Bible," Ancestry.co.uk, 30 May 1999, http://www.boards.ancestry.co.uk/localities .northam.usa.states.tennessee.counties.greene/2572/mb.ashx.

29. Noe, *Southwest Virginia's Railroad*, 67–108; Hsiung, *Two Worlds*.

30. John Harrington, "He Spoke: Biography and Scientific Work of Richard Lynch Garner," Biographer's Notes folder, box 5, Richard Lynch Garner Papers,

National Anthropological Archives, Smithsonian Institution, Washington, D.C. (hereafter cited as RLGP) (Washington, D.C.: Government Printing Office, 1941), 7–8.

31. On Garner's military career, see Microfilm N598-39, 14 December 1864, Selected Records of the War Department Related to Confederate Prisoners of War 1861–1865, U.S. War Department, National Archives and Records Administration, Washington, D.C. (hereafter cited as Selected Records); Microfilm N598-92, 8 January 1865, Selected Records; Microfilm N598-92, 17 February 1865, Selected Records; Richard Lynch Garner to Charles Furlong, 2 December 1916, "Lot 267," artfact.com, accessed 23 January 2010, www.artfact.com/auction-lot/still-on -parole-in-1916-1-c-lsmvzhttbd; Harrington, "He Spoke," 8–9.

32. Noe, *Southwest Virginia's Railroad*, 109–138; McKnight, *Contested Borderland*; Larry Gordon, *Last Confederate General*, 67–140.

33. Radick, *Simian Tongue*, 88.

34. Garner, *Autobiography of a Boy*, 86.

35. Details on Garner's teaching career are very sparse. For the reference to Williamsburg, Kentucky, see "Professor Garner, Explorer, Dies at 72," *New York Times*, 24 January 1920, 11. The Whitely County Historical Society in Williamsburg has no reference to Garner in their records.

36. Hugh Gilmore to Walter Hendricks, 18 February 1990, Richard Lynch Garner File, Washington County Historical Society, Abingdon, Va.

37. Dotson, *Roanoke, Virginia*.

38. Radick, *Simian Tongue*, 88–90.

39. Richard Garner, "Nancy Bet and Other Stories," n.d., ca. 1890, Nancy Bet folder, box 3, RLGP.

40. Algeo, "Locals on Local Color."

41. Lyon, *Success Story*, 54–109.

CHAPTER ONE. The Southern Gabonese Coast in the Age of Garner

1. Savage and Wyman, "Notes of the External Characters," in Burham, *Science in America*, 120.

2. For a broad overview of primate trading in Gabon, see Rich, "Chimpanzees in the Colonial Maelstrom."

3. Garner, "Adventures in Central Africa: Domestic Arrangements," 598.

4. The best overview of the ecology of the southern Gabonese coast is Vande weghe, *Les parcs nationaux du Gabon*.

5. Du Chaillu, *Explorations and Adventures, Journey to Ashango-Land*. On Du Chaillu's career in Gabon, see Patterson, "Paul B. Du Chaillu"; Mandelstam, "Du

Chaillu's Stuffed Gorillas"; McCook, "It May Be Truth"; and Gaulme, "Paul Du Chaillu," in Hombert and Perrois, *Coeur d'Afrique*. The Hombert and Perrois study is by far the best work on Du Chaillu.

6. "Accidents: Mr. Frank James, Killed by a Wounded Elephant, while Hunting on the Gaboon," *Times of London*, 3 May 1890, 7; 2 August 1897 entry, Journaux de la Communauté de Sainte Anne de Fernan Vaz, 1888–1939, Archives of the Congregation of the Holy Spirit, Chevilly Larue, France (hereafter cited as JC). On German visitors to Fernan Vaz such as the German noble Hugo von Koppenfels, see Brehm, *Tierleben*, 651.

7. The best source on the environment in this region is Rabenkogo, "Le littoral du Nkomi (Gabon)."

8. At least in 1908, Garner noted his lack of fluency in Omyènè. Richard Garner, "Heathen Rites," n.d., ca. 1908, Habits of Wild Animals folder, box 2, RLGP, 10.

9. Gray, *Colonial Rule and Crisis*.

10. On the impact of the slave trade on Fernan Vaz and Nkomi in the nineteenth century, see Agondjo-Okawe, "Structures parentales gabonaises et développement," 120–123; Gaulme, *Le pays de Cama*; and Ambouroue-Avaro, *Un peuple Gabonais à l'aube*, 175–211.

11. On the *rengondo* Oyembo Onanga, see Agondjo-Okawe, "Structures parentales gabonaises et développement," 30–40; Gaulme, *Le pays de Cama*, 140–150, 185–195; and Ambouroue-Avaro, *Un peuple Gabonais à l'aube*, 175–211. His violent rule and his wealth remain important in the historical imagination of Fernan Vaz, as testified by his looming presence in twenty interviews made in Fernan Vaz in December 2008 and January 2009.

12. For a more detailed discussion of these arrangements, see chapter 6 and Rich, "White Coronations and Magical Boycotts."

13. Ambouroue-Avaro, *Un peuple Gabonais à l'aube*, 194–200.

14. On the Orungu, see Patterson, *Northern Gabonese Coast to 1875*; Ambouroue-Avaro, *Un peuple Gabonais à l'aube*, 144–156; and M'Bokolo, *Noirs et blancs*.

15. On Gisir-speaking communities, see Gray, *Colonial Rule and Crisis*, 116–123, 146–158. Gisir communities, now generally known as the Eshira ethnic group, have unfortunately been neglected in scholarly research in Gabon. On Fang societies in Gabon during the late nineteenth and early twentieth century, see Chamberlin, "Competition and Conflict"; Fernandez, *Bwiti*; Cinnamon, "Long March of the Fang"; Bernault, "Dévoreurs de la nation"; Jean-Baptiste, "Une ville libre?"; and Cadet, *Histoire des Fang*.

16. These small groups have been entirely ignored in Gabonese scholarship since the early twentieth-century writings of Gabonese priest and intellectual

André Raponda Walker. See *Notes d'histoire du Gabon*, 153–155; and Aleko and Peuch, "Notes sur la lagune Ngovè."

17. Aschemeier, "Life among the Natives," 22; Ambrose Means, "Means in Africa," n.d., Ambrose Means Papers, private collection, Cathy Carter, Prescott, Ariz., 1–4. Means visited Sette Cama and Fernan Vaz in the spring of 1914. The essay is held by Linda Thompson, Means's granddaughter, who kindly sent the author a copy.

18. On gender roles in Nkomi society, see Agondjo-Okawe, "Structures parentales gabonaises et développement," 78–104, 123–130.

19. Richard Garner, "Marooned by Savages," n.d., ca. 1905, Man as He Will Be folder, box 3, RLGP, 2.

20. Simonton, "The French Congo," *Africa Mail*, 13 August 1909, 444; 8 October 1909, 3.

21. Richard Garner, 5 October 1905 entry, Diary 1905–1906 folder, box 1, RLGP.

22. Richard Garner, "Buiti," n.d., Buiti folder, box 1, RLGP, 12–13.

23. Richard Lynch Garner, *Autobiography of a Boy*, 8–11, 56–58. This manuscript was compiled from a series of Garner's letters written in 1904.

24. Albert Veistroffer, "Rapport sur la Fernan Vaz," 5 February 1898, Rapports Politiques, 1896–1920, Gouvernement Général de l'Afrique Équatoriale Française, série 4(1)D1, Centre d'Archives d'Outre-mer, Aix-en-Provence, France (hereafter cited as RP).

25. 2 and 18 August 1897 entries, microfilm T2 B5, JC; Fôret, "Le lac Fernan Vaz"; Veistroffer, "Rapport sur la Fernan Vaz"; Gaulme, "Un problème d'histoire du Gabon."

26. Veistroffer, *Vingt ans*, 213, 217.

27. Gaulme, "Un problème d'histoire du Gabon."

28. Madeline Yeno, interview with author, 27 December 2008, Sainte Anne, Gabon.

29. On concessionary companies in Gabon between 1899 and the 1920s, see Coquery-Vidrovitch, *Le Congo*, 238–257, and Gray, *Colonial Rule and Crisis*, 151–163.

30. "Communauté de Sainte Anne," *Bulletin de la Congregation*.

31. Cercus, "Rapport politique: Fernan Vaz et Sette Cama," March 1907, série 4(1)D3, RP.

32. Nze Ndong, Marie Nyi Nyondome, and Jeanne Eloghe, interview with author, 30 December 2008, Kongo, Gabon.

33. Jean-Claude Dikoumba, interview with author, 21 December 2008, Ndougou, Gabon; Rich, "Cruel Guards and Anxious Chiefs"; Mikadi and Pascal Nkundi, interview with author, 19 July 2007, Biaka, Gabon.

34. Henriette Nkourou Ratowo, interview with author, 23 December 2008, Ombouè, Gabon.

35. Cercus, "Rapport d'une tournée d'inspection," 2–7 August 1906, série 8Y24, RP, 45.

36. Simonton, "French Congo," 23 August 1909, 23.

37. Ibid., 9 July 1909, 393.

38. Reade, *Savage Africa*, 176.

39. Simonton, "French Congo," 13 August 1909, 444.

40. Richard Garner, "An American Lawgiver: Captain Lawler," n.d., ca. 1906, Africa and the World War folder, box 1, RLGP.

41. Richard Garner to William Hornaday, 7 May 1913, box 53, Correspondence R. L. Garner, Director's Office, William Hornaday Papers, Wildlife Conservation Society Library, New York (hereafter cited as WHP).

42. Garner, "Religion of African Cannibals," 305.

43. Ditmars, *Confessions of a Scientist*, 226.

44. On the impact of World War I on Gabon, see Loungou Mouele, "Le Gabon,"; Dubois, "Le prix d'une guerre"; Gray, *Colonial Rule and Crisis*, 153–161; and Rich, *Workman Is Worthy*, 64–85.

45. On Charmanade's brutality and the misfortunes of Gabonese people in Fernan Vaz during the war, see 31 October 1916, 30 November 1916, 15 January 1917, 10 May 1917, and 21 September 1918 entries, microfilm T2 B6, JC.

46. Charles Aschemeier to W. D. Ravenel, 24 October 1918, folder 12, record unit 45, Collins-Garner Expedition, Smithsonian Institution Archives, Washington, D.C. (hereafter cited as CGE).

47. Richard Garner, "Famine and the Food Problem in Africa," n.d., ca. 1918, Facial Expression folder, box 2, RLGP.

48. Rich, "Gabonese Men for French Decency."

49. Rabenkogo, "Le littoral du Nkomi (Gabon)," 205–213.

50. On conservation in the Fernan Vaz region in the early twenty-first century, see Rabenkogo, "Le littoral du Nkomi (Gabon)," 221–260.

51. For some important works on Leopold II's Congo, see Slade, *King Leopold's Congo*; Hochschild, *King Leopold's Ghost*; Ó Síocháin and O'Sullivan, *Eyes of Another Race*; Dumoulin, *Léopold II, un roi génocidaire?*; and Vanthemsche, *La Belgique et le Congo*.

52. Morel, *British Case in French Congo*; Challaye, *Le Congo français*; Cookey, "Concession Policy"; Coquery-Vidrovitch, *Le Congo*; Jean Martin, *Savorgnan de Brazza*.

53. Caudill, *Night Comes to the Cumberlands*; Lewis, Johnson, and Askins, *Colonialism in Modern America*; Eller, *Miners, Millhands, and Mountaineers*.

54. For a brief discussion of how academics became disillusioned with the colonization model, see Billings, Pudup, and Waller, introduction to *Appalachia in the Making*, 9–15; and Billings and Blee, *Road to Poverty*, 11–14.

55. Rich, "King or Knave"; Rich, "Civilized Attire."

56. Algeo, "Locals on Local Color"; West, *From Yeoman to Redneck*.

CHAPTER TWO. Garner's Animal Business in Africa and America

1. On the rise of zoos and their links to colonial expansion in the nineteenth century, see Croke, *Modern Ark*, 140–155; Rothfels, *Savages and Beasts*; Hancocks, *Different Nature*, 33–54; Baratay and Hardouin-Fugier, *Zoo*, 113–130; Dehler, "American Crusader"; Hanson, *Animal Attractions*; and Murray, "Lives of the Zoo."

2. Haraway, *Primate Visions*, 26–58.

3. Brechin, "Conserving the Race"; Stern, *Eugenic Nation*; Spiro, *Defending the Master Race*.

4. On animal dealers, see Brandon-Jones, "Edward Blyth"; Rothfels, *Savages and Beasts*, 44–80; Hanson, *Animal Attractions*, 71–99; and Jones, "Gorilla Trails in Paradise."

5. Rothfels, *Savages and Beasts*, 44–80; Ames, *Carl Hagenbeck's Empire*.

6. Jacobs, "Intimate Politics of Ornithology."

7. "Museum Donations," *Calendar of the University of Toronto and University College*, 1896–1897 (Toronto: University of Toronto, 1897), 28.

8. Richard Garner to Harry Garner, 7 February 1905, Outgoing Letters folder, box 1, RLGP.

9. On concessionary companies in Gabon, see Coquery-Vidrovitch, *Le Congo*.

10. Cercus, "Rapport d'une tournée d'inspection," 5–6.

11. 11 March 1903 entry, microfilm T2 B5, Journaux de la Communauté; Cercus, "Rapport sur la Compagnie du Fernan Vaz," 2–7 August 1906, séric 8Y24, RP, 45; Richard Garner, "The French Congo," n.d., ca. 1908, Facial Expression folder, box 2, RLGP; Simonton, "French Congo," 13 August 1909, 443–444; 20 August 1909, 453–455.

12. Richard Garner to Mangafa, 2 August 1907, Financial Records folder, box 5, RLGP.

13. Richard Garner, "Specimen Letters of Natives," n.d., ca. 1906, Sanctuary to Women folder, box 4, RLGP.

14. Garner, "Marooned by Savages."

15. Richard Garner to Harry E. Garner, 7 February 1905, Outgoing Letters folder, box 1, RLGP.

16. Garner, 6 March 1905 entry, Diary 1905–1906, RLGP; Garner to Hornaday, 14 January 1912, box 52, WHP; Garner, "Adventures in Central Africa: Domestic Arrangements."

17. Garner to Hornaday, 14 January 1912, box 52, WHP.

18. Garner, "In the Clutches of Cannibals."

19. Garner, 2 January 1905, 10 January 1905, 11 February 1905, 23 March 1905, 25 October 1905, and 29 October 1905 entries, Diary 1905–1906, RLGP.

20. Richard Garner to Harry E. Garner, 18 May 1909 and 18 August 1904, RLGP.

21. Ibid., 14 June 1904.

22. Richard Garner to Professor Holmes, 6 December 1900, Outgoing Letters folder, box 1, RLGP.

23. "Monkey Intelligence Said to Surpass That of Many Natives," *Syracuse Evening Herald*, 17 February 1896, 2.

24. Garner to Holmes, 6 December 1900, RLGP.

25. Hornaday, "Bird's-Eye View," 232.

26. Richard Garner to Harry E. Garner, 20 May 1906, RLGP.

27. Bridges, *Gathering of Animals*, 346–347.

28. Garner to Hornaday, 27 February 1911, box 47, WHP.

29. Garner to Hornaday, 6 April 1911, box 47, WHP; Garner to New York Zoological Park, 11 April 1911, box 53, WHP.

30. Hornaday to Garner, April 1911, and Garner to Hornaday, 29 April 1911, box 47, WHP.

31. Although Garner and Imbrie coauthored a description of their visit to Gabon, this text apparently has not survived.

32. Garner to Hornaday, 20 May 1911, box 47, WHP. Similar comments appear in Garner to Hornaday, 13 and 18 July 1911.

33. Ibid., 13 and 18 July 1911.

34. "Miss Fatima Arrives," *San Antonio Light*, 8 October 1911, 27; "Gorilla Has Grouch," *Mills County Tribune*, 13 November 1911, 4.

35. Bridges, *Gathering of Animals*, 347–348.

36. Garner to Hornaday, 18 August 1912, box 47, and 4 January 1913, box 52, WHP.

37. Hornaday to Garner, 28 January 1914 and 15 April 1914, box 52, WHP.

38. Garner to Hornaday, 4 January 1913 and 9 February 1914, box 52, WHP.

39. Ibid., 29 September 1913 and 9 February 1914.

40. Ibid., 7 May 1913.

41. Ibid., 31 May 1913, 31 August 1913, and 29 September 1913.

42. Ibid., 31 August 1913 and 1 January 1914.

43. On Jones's expedition to Gabon, see Easton and Brown, *Lord of Beasts*, 201–229.

44. Hornaday to Garner, 28 January 1914, box 52, WHP.

45. Garner to Hornaday, 6 and 17 June 1914, box 52, WHP; Easton and Brown, *Lord of Beasts*, 212–223.

46. Easton and Brown, *Lord of Beasts*, 212–213; Trial, *Okoume*.

47. Ambrose Means, "Observations: The Gorilla in Africa," 1914, Ambrose Means Papers. This short essay is presently in the possession of Cheryl McAdams, one of Mr. Means's granddaughters.

48. Hornaday to Garner, 15 March 1913 and 9 September 1913, box 52, WHP.

49. Garner to Hornaday, 1 January 1914, box 52, WHP.

50. Hornaday to Garner, 15 April 1914, box 52, WHP.

51. Blathwayt, "Wild Animals," 29; Hornaday to Garner, 1 October 1914, box 52, WHP.

52. Garner to Hornaday, 27 June 1914, box 52, WHP.

53. Ditmars, *Confessions of a Scientist*, 226.

54. Hornaday, "Department of Mammals."

55. On World War I in Gabon, see Loungou Mouele, "Le Gabon"; Dubois, "Le prix d'une guerre"; and Gray, *Colonial Rule and Crisis*, 153–161.

56. Neither officials nor missionaries stated the exact nature of Charmanade's charges, and his personal file in the French colonial archives remains unavailable to researchers. 22 August 1916, 6 November 1916, 31 October 1916, 30 November 1916, 15 January 1917, 10 May 1917, and 21 September 1918 entries, microfilm T2 B6, JC.

57. "Museum Notes."

58. William Hornaday to Charles Wolcott, 16 December 1916, folder 9, CGE.

59. Henry Fairfield Osborn to Charles Wolcott, 8 December 1916; Hornaday to Wolcott, 16 December 1916, CGE.

60. Leonhard Stejneger to Richard Rathbun, 26 November 1915, folder 8, CGE.

61. Richard Garner to Charles Wolcott, 1 December 1915, folder 8, CGE.

62. Charles Wolcott to J. H. Schiff, 19 November 1915, folder 8, CGE.

63. Alfred Collins to Charles Wolcott, 23 October 1916; [illegible name] to Charles Wolcott, 26 October 1916; Collins and Wolcott to State Department Secretary, 22 November 1916, folder 9, CGE.

64. Wolcott to Ambassador Jean-Jules Jusserand, 6 and 12 December 1916, and Jusserand to Wolcott, 16 December 1916 folder 9, CGE.

65. Aschemeier to Miller, 18 January 1917, folder 10, CGE.

66. Wolcott to Collins, 5 June 1917, and R. Rathburn to Charles Aschemeier, 14 July 1917, folder 10, CGE.

67. Garner to Wolcott, 20 March 1917 and 25 May 1917, folder 10, CGE; Garner, "Adventures in Central Africa: Domestic Arrangements," 596.

68. Garner, "Adventures in Central Africa: Domestic Arrangements," 599.

69. Rand, Friedmann, and Traylor, "Birds from Gabon," 228–229. Aschemeier had provided the authors with information on village and place names as well as on Garner's health. Unfortunately, Aschemeier's private papers appear to have been lost after his death in 1973, and it is unclear if the information in this article came from his memory or his notes.

70. Garner to Wolcott, 6 August 1917, folder 11, CGE.

71. Aschemeier to Wolcott, 9 February 1919, folder 12, CGE. The record book, which gives a detailed list of birds collected each day, is available at the Smithsonian Institute (Charles Aschemeier, Field Notes 1917–1918, record unit 105, Smithsonian Institution Archives, Washington, D.C.). Unfortunately, the online Smithsonian catalogue has the list incorrectly filed in the Botany Department archives.

72. Aschemeier to Miller, Collins to Gerritt Miller, 26 December 1917, folder 10, CGE; Garner, "Famine and the Food Problem," 3–5.

73. Garner, "Adventures in Central Africa: Domestic Arrangements," 601–602.

74. Ibid., 600, 604.

75. Garner, "Adventures in Central Africa: Peculiar Monkeys," 847, 850.

76. Alfred Collins to Harry E. Garner, 24 February 1920, Incoming Letters folder, box 1, RLGP.

77. Aschemeier, "Gorilla at Home," 19.

78. Bernault, "Body, Power and Sacrifice"; Gray, *Colonial Rule and Crisis*, 195–224.

79. Aschemeier, "Life among the Natives," 21. On leopards raising their young, see Rand, Friedmann, and Traylor, "Birds from Gabon," 229.

80. Raponda Walker, *Notes d'histoire du Gabon*, 154–155.

81. Circonscription Nkomis, June and December 1910, série 4(1)D5, RP; Nkomis, June 1912, série 4(1)D9, RP.

82. Aschemeier to Wolcott, 9 February 1919, folder 12, and 19 October 1918, folder 11, CGE.

83. Garret Miller to Stejneger, 23 October 1918, folder 11, CGE.

84. Agent Général of Chargers Réunis to Charles Wolcott, 19 November 1918, folder 11, CGE.

85. Garner to Wolcott, 12 April 1918, folder 11, CGE.

CHAPTER THREE.   Is the Monkey Man Manly Enough?

1. Rotundo, *American Manhood*; Bederman, *Manliness and Civilization*; Hoganson, *Fighting for American Manhood*; Kasson, *Perfect Man*; Pettegrew, *Brutes in Suits*.

2. Friend and Glover, *Southern Manhood*.

3. Gilmore, *Gender and Jim Crow*, 65.

4. Friend, "From Southern Manhood," in Friend, *Southern Masculinity*, xv.

5. West, *From Yeoman to Redneck*.

6. Garner, *Autobiography of a Boy*, 60, 78–81. Subsequent citations to this work are given in the text.

7. Renoff, *Big Tent*, 11–32.

8. Garner, 15 October 1905 entry, Diary 1905–1906, RLGP.

9. For general works on atheism in America in the Gilded Age and the Progressive Era, see Sidney Warren, *American Freethought*; Turner, *Without God, without Creed*; and Jacoby, *Freethinkers*. On atheists outside of the East Coast in the United States, see Whitehead and Muhrer, introduction to *Freethought on the American Frontier*, 15–23, and Jacoby, *Freethinkers*, 174–178, 210–212.

10. On Ingersoll, see Orvin Larson, *American Infidel*.

11. Garner to Holmes, 6 July 1900, RLGP.

12. Richard Garner to Harry Garner, 16 July 1909, RLGP.

13. Sondage, *Born Losers*.

14. Frank Baker to Dr. Langley, 3 May 1892, folder 25, box 12, record unit 365, National Zoological Park, Smithsonian Institution Archives, Washington, D.C. (hereafter cited as NZP).

15. Frank Baker to W. K. Brooks, 22 April 1892, box 12, NZP.

16. Herzig, *Suffering for Science*.

17. Jones, "Gorilla Trails in Paradise."

18. Garner, *Speech of Monkeys*, v; Nettels, *Language, Race, and Social Class*, 198.

19. "Mr. Garner and the Monkeys," *New York Times*, 17 February 1892, 2.

20. Garner, "What I Expect to Do," 713, 716, 717.

21. Veistroffer, *Vingt ans*, 217.

22. "Dr. Garner and Monkey Language," *Salt Lake Tribune*, 21 April 1895, 5.

23. Smith, *Glimpses of Africa*, 204.

24. Richard Garner to Harry Garner, 21 June 1907, RLGP.

25. Rich, "Une Babylone Noire"; Jean-Baptiste, "Une ville libre."

26. Veistroffer, *Vingt ans*, 64.

27. Photographs folder, box 5, RLGP.

28. On concessionary companies in Gabon, see Coquery-Vidrovitch, *Le Congo*. For examples of Garner's close ties to colonial officials and CCFV employees, see Garner, 6 March 1905 entry, Diary 1905–1906, RLGP; Garner to Hornaday, 14 January 1912, box 52, WHP; Garner, "Adventures in Central Africa: Domestic Arrangements," 597.

29. Radick, *Simian Tongue*, 138–154.

30. "Weekly Gazette," *Fort Wayne Journal Gazette*, 19 November 1896, 4.

31. "Town Topics," 13.

32. Garner to Holmes, 6 December 1900, RLGP. Garner underlined the words.

33. Richard Garner to Harry Garner, 24 October 1906, RLGP.

34. Ibid., 21 June 1907, 4.

35. Ibid., 27 March 1907 and 21 June 1907, 4.

36. The following is drawn from Richard Garner, "Things That Must Come to Pass: The Rights of Franchise under the Constitution of the United States; Political Prophesies," 1908, Tails of Dogs and Other Animals folder, box 4, RLGP, 9–13, 11.

37. Richard Garner to Harry Garner, 21 June 1907, RLGP, 17.

38. For discussions of racial degeneration in colonial Africa, see Kennedy, *Islands of White*, and Newell, *Forger's Tale*.

39. Richard Garner to Harry Garner, 21 June 1907, RLGP, 13, 22.

40. Ibid., 25, 26.

41. Ibid., 21 June 1907 and 11 November 1907, 38.

42. Ibid., 21 June 1907, 28. Subsequent citations to this work are given in the text.

43. Ibid., 24 May 1907.

44. Ibid., 18 January 1907, 24 May 1907, 11 November 1907, 28 April 1908, and 1 September 1908.

45. Garner, "What Next?," n.d., ca. 1908, What Next? folder, box 4, RLGP.

46. Garner, "A State of Mind and a State of Facts," n.d., ca. 1908, Spots of Earth folder, box 4, RLGP.

47. "A Monkey-World, with Generals, Society Belles and Many Diversions," *Atlanta Journal-Constitution*, 24 August 1902, D3; Garner, "Big Game Hunting," 400.

48. Garner, "Cannibals," 597–603.

49. "Knows All Monkey Talk," *New York Tribune*, 22 August 1910, 2.

50. "Garner's Monkey," 20.

51. Ibid., 2 June 1911.

52. Richard Garner to Dr. Miller, n.d., ca. November 1915, folder 9, CGE.

53. Hornaday, *Minds and Manners*, 80.

54. "Old Traveler Claims Jungle Is Safer Than Streets of City," *Pittsburgh Post-Gazette*, 29 July 1919, 19.

55. "Current Notes," 198.

56. Radick, *Simian Tongue*, 172.

57. Hornaday to Wolcott, 16 December 1916, folder 9, CGE.

58. Richard Garner to Charles Furlong, 2 January 1917, "Lot 267,"

artfact.com, accessed 31 January 2010, http://www.artfact.com/auction-lot
/confederate-letters-documents-former-confed-1-c-z6g2m33oyf.

59. Yerkes and Yerkes, *Great Apes*, 164.

CHAPTER FOUR. Race, Knowledge, and Colonialism
in Garner's African Writings

1. Garner, "Things That Must Come," RLGP, 13.

2. Duignan and Gann, *United States and Africa*; Hickey and Wylie, *Enchanting Darkness*; Bederman, *Manliness and Civilization*, 207–213; Jacobson, *Barbarian Virtues*, 105–117; Kasson, *Perfect Man*; Mayer, *Artificial Africas*.

3. Stecopoulos, *Reconstructing the World*, 18–38.

4. Lake and Reynolds, *Drawing the Global Colour Line*.

5. Love, *Race over Empire*.

6. Silber, *Romance of Reunion*.

7. Garner, *Autobiography of a Boy*, 11, 56–57.

8. Noe, *Southwest Virginia's Railroad*, 67–84; Dunaway, *Slavery in the Mountain South*, 117–118.

9. Summers, *History of Southwestern Virginia*, 522–523.

10. For details on the Civil War in and around Abingdon, see Noe, *Southwest Virginia's Railroad*, 109–138; McKnight, *Contested Borderland*, 188–226; and Larry Gordon, *Last Confederate General*, 67–86, 120–145.

11. On Saltville, see Mays, *Saltville Massacre*; McKnight, *Contested Borderland*, 209–213; and Larry Gordon, *Last Confederate General*, 126–127.

12. Richard Garner, "Feast of Fire," 1889, Facial Expression folder, box 2, RLGP.

13. Garner, "The Indian and His Dog," n.d., The Indian and His Dog folder, box 2, RLGP.

14. Garner, "Caesar's Ghost," n.d., Can Monkeys Talk? folder, box 1, RLGP.

15. Garner, "The Negro Problem," 1905, Nancy Bet folder, box 3, RLGP, 6, 8.

16. Garner, "Negro Question," Nancy Bet folder, box 3, RLGP, 9, 10.

17. Garner, *Speech of Monkeys*, 14, 16, 152, 96.

18. Garner, *Apes and Monkeys*, 91.

19. Richard Garner to Harry Garner, 1 September 1908, RLGP.

20. Garner, "Caucasian Monkeys," Can Monkeys Talk? folder, box 1, RLGP.

21. Cited in Fishel and Quarles, *Black American*, 379–381.

22. Tillinghast, *Negro in Africa and America*, 26.

23. Garner, "The Negro," n.d., ca. 1905, Nancy Bet folder, box 3, RLGP, 1.

24. Garner, "The Moon's Fire," n.d., Monkeyland folder, box 3, RLGP, 5.

25. Garner, "Native Institutions," 370, 375.

26. Garner, "Negro Loyalty," n.d., Nancy Bet folder, box 2, RLGP.

27. Garner, "Marooned by Savages," 1.

28. Garner, "Negro," 1.

29. Garner, "Religion of African Cannibals," 313.

30. Garner, "Heathen Prayers," n.d., ca. 1906, The Habits of Wild Animals folder, box 2, RLGP, 7.

31. Garner, "Native Institutions," 378, 372.

32. Garner, "Religion of African Cannibals," 314.

33. Garner, 19 November 1905 entry, Diary 1905–1906, RLGP.

34. Garner, "Sanctuary to Women," n.d., ca 1906, Sanctuary to Women folder, box 4, RLGP.

35. Garner, "Superstitions of the West African Tribes," n.d., Spots on Earth folder, box 4, RLGP, 8.

36. Garner, "The Phonograph among the Savages," n.d., Phonograph among the Savages folder, box 3, RLGP; Garner, "Witch Rain," n.d., ca. 1906, What Next? folder, box 4, RLGP.

37. "Fetish Tusk," accession no. 36893, catalog no. 206111, Department of Anthropology, Smithsonian Institution, accessed 25 February 2010, http://www.collections.nmnh.si.edu/anth/pages/nmnh/anth/imagedisplay.php?irn=8015146&reftable=enmnh&refirn=8357585.

38. Garner, "Heathen Prayers to Heathen Gods," n.d., ca. 1907, Habits of Wild Animals folder, box 1, RLGP.

39. Garner, "Adventures of Central Africa: Domestic Arrangements," 601.

40. Garner, "Educated Natives," n.d., ca. 1905, Educated Animals folder, box 2, RLGP, 1.

41. Garner, "Heathen and the Bible," n.d., Habits of Wild Animals folder, box 2, RLGP.

42. Garner, "Missions and Missionaries," n.d., Man as He Will Be folder, box 3, RLGP, 3.

43. Garner, *Autobiography of a Boy*, 68.

44. Garner, "Negro Problem," 3–5.

45. Guterl, *Color of Race in America*, 27–61.

46. Garner, "The French Congo," n.d., ca. 1908, Facial Expression folder, box 2, RLGP, 1. Subsequent citations to this work are given in the text.

47. Garner, "Colonial Officials" and "Colonizing Africa, " n.d., ca. 1906, Civilized Savagery folder, box 2, RLGP.

48. Jean-Baptiste, "Une ville libre?"

49. Morel, *British Case in French Congo*, 146.

50. Welch, *Response to Imperialism*, 101–112.

51. Garner, "Negro Problem," 10, 13.

52. Fredrickson, *Black Image*, 229, 263–266.

53. Richard Garner to Harry Garner, 21 June 1907, RLGP.

54. The Smithsonian Institution has placed all the surviving lantern slides online. National Anthropological Archives, lot 81-58a, accessed 23 February 2010, http://www.collections.si.edu/search/slideshow_embedded?xml="http://sirismm .si.edu/naa/viewer/PhotoLot81-58a_Gallery/PhotoLot81-58a.xml.

55. Garner, "Adventures in Central Africa: Domestic Arrangements," 597.

CHAPTER FIVE. African Animals for White Supremacy

1. Garner, *Apes and Monkeys*, 12.

2. Thomas, *Man and the Natural World*, 101–109, 173–191; Ritvo, *Animal Estate*, 85–115; Grier, *Pets in America*, 133–142.

3. Rothfels, "How the Caged Bird Sings," in Kete, *Cultural History of Animals*, 111; Murray, "Lives of the Zoo."

4. Brantz, "Domestication of Empire," in Kete, *Cultural History of Animals*, 75.

5. MacKenzie, *Empire of Nature*; Haraway, *Primate Visions*, 26–58; Storey, "Big Cats and Imperialism"; Bederman, *Manliness and Civilization*, 207–213; Neumann, *Imposing Wilderness*, 98–143; Steinhart, *Black Poachers, White Hunters*.

6. A good introduction to this literature is Van Sittert and Swart, *Canis Familiaris*.

7. Kete, introduction to *Cultural History of Animals*, 15.

8. Garner, *Speech of Monkeys*, 53, 54, 56.

9. Garner, *Apes and Monkeys*, 105. Subsequent citations to this work are given in the text.

10. Richard Garner to Harry Garner, 21 June 1907, RLGP.

11. Garner, "Just as It Happened: A True Unvarnished Statement of Events in the Life and Experiences of R. L. Garner," n.d., ca. 1908, Just as It Happened folder, box 2, RLGP, 13.

12. Garner, "Hermit's Home," 110, 120.

13. Garner, "Just as It Happened," 13.

14. Garner, "Dinkie and Dot: The Story of Two Orphan Monkeys," n.d., Dangers in Traveling in Africa folder, box 2, RLGP, 1.

15. Ibid., 2.

16. Garner, "In an African Jungle," *Hartford Courant*, 20 June 1906, 10.

17. Garner, "Contemporary Ancestors of Ours," 60.

18. Garner, Diary 1905–1906, RLGP.

19. "Intelligent Chimpanzee," 750.

20. "Can Monkeys Talk?," 492.

21. Preston, "Educating a Chimpanzee."

22. Garner, "Why I Know," 23.

23. Garner, "Contemporary Ancestors of Ours," 77.

24. Grier, *Pets in America*.

25. Rich, "Une Babylone Noire"; Jean-Baptiste, "Une ville libre." This situation appears very similar to white society in Brazzaville before 1914. See Phyllis Martin, *Leisure and Society*, 187–194.

26. Rich, "Civilized Attire."

27. Sinha, *Colonial Masculinity*.

28. Prestholdt, *Domesticating the World*, 117–146, 147–170.

29. Raponda Walker, *Contes Gabonaises*.

30. Jean-Robert Ngomba, interview with author, 27 December 2008, Sainte Anne, Gabon.

31. Garner, "Bubu: A Faithful Dog," 1908, Base Ingratitude folder, box 1, RLGP, 1, 3.

32. Cheang, "Women, Pets, and Imperialism"; Skabelund, "Breeding Racism"; Van Sittert and Swart, *Canis Familiaris*; Skabelund, "Fascism's Furry Friends."

33. Garner, "Bubu," 16. Subsequent citations to this work are given in the text.

34. Rich, "After the Slave Trade."

35. Tropp, "Meaning of Colonial Intervention"; Richard Gordon, "Fido."

36. Richard Lynch Garner, "Specimens of Native Letters," n.d., Sanctuary to Women folder, box 4, RLGP.

37. Patterson, "Vanishing Mpongwe."

38. Fabian, Out of Our Minds, 101.

39. Brockington, Fortress Conservation; Fay and Nicholas, Last Place on Earth; Garland, "Elephant in the Room," 54–58.

CHAPTER SIX. An American Sorcerer in Colonial Gabon

1. Garner, "Witch Rain."

2. Ibid., 2.

3. Sillans and Raponda Walker, *Rites et croyances*, 22–23.

4. Garner, "Witch Rain," 5–6.

5. Ibid., 2.

6. Fabian, *Out of Our Minds*, 265–270.

7. As noted by Adas, *Machines as the Measure*.

8. Ragonondo to Garner, 21 June 1907, Outgoing Letters folder, box 1, RLGP.

9. Cookey, "Concession Policy"; Dupré, *Un ordre et sa destruction*; Coquery-

Vidrovitch, *Le Congo*; Cinnamon, "Long March of the Fang"; Gray, *Colonial Rule and Crisis*, 141–161.

10. Vansina, *Paths in the Rainforest*, 239; Guyer and Eno Belinga, "Wealth in People"; Gaulme, *Le pays de Cama*, 251–252.

11. Bernault, *Démocraties ambigües*; Cinnamon, "Long March of the Fang"; Gray, *Colonial Rule and Crisis*, 195–224; Bernault, "Body, Power, and Sacrifice."

12. Cooper, *Colonialism in Question*, 185–186, 215.

13. Geschiere, *Modernity of Witchcraft*; Tonda, *Le Souverain modern*; Bernault, "Body, Power, and Sacrifice."

14. Garner, "Phonograph among the Savages," 8.

15. Ambouroue Avaro, *Un peuple Gabonais à l'aube*, 175–211.

16. Guyer and Eno Belinga, "Wealth in People."

17. Appadurai, "Commodities," 21. In "Tournaments of Value," Peter Geschiere has expanded on this discussion to explore the concept's value in understanding Beti responses to economic and political changes between 1890 and 1914 in Cameroon.

18. Bonhomme, "Les tribulations de l'esprit blanc," 500–501.

19. Sillans and Raponda Walker, *Rites et croyances*, 79–96.

20. MacGaffey, *Kongo Political Culture*, 217.

21. Sillans and Raponda Walker, *Rites et croyances*, 87, 100–101.

22. Agondjo-Okawe, "Structures parentales gabonaises et développement," 21–30.

23. For a general overview of the early history of *bwiti* among Fang communities, see Fernandez, *Bwiti*; Swiderski, *La religion Bouiti*; Mary, *Le défi du syncrétisme*; and Bonhomme, *Le miroir et la crâne*. On *bwiti* ceremonies practiced by Nkomi people, see Gaulme, "Le Bwiti chez les Nkomi."

24. 11 October 1897 entry, microfilm T2 B5, JC.

25. Garner, "Phonograph among the Savages," 2, 3.

26. Ibid., 4.

27. Ibid., 8.

28. Bernault, "Body, Power, and Sacrifice," 213–216, 233–237.

29. Garner, "Phonograph among the Savages," 6.

30. Laburthe-Tolra, *Les seigneurs de la forêt*, 318–327; Cinnamon, "Long March of the Fang," 173–181; Rich, "My Matrimonial Bureau."

31. Garner, "Phonograph among the Savages," 10.

32. The following account is drawn from Garner, "Marooned by Savages," 1.

33. Garner, "Memo of Money Lent and Dashes Given to Anjanga," n.d., ca. 1905, Financial Reports folder, box 5, RLGP.

34. Agondjo-Okawe, "Les droits fonciers coutumiers," 1146.

35. Garner, "Marooned by Savages," 3.

36. Agondjo-Okawe, "Les droits fonciers coutumiers," 1145.

37. Agondjo-Okawe, "Structures parentales gabonaises et développement," 144–146.

38. Ibid., 3, 4.

39. Ibid., 5.

40. Garner, "Among the Cannibals," n.d., Africa and the World War folder, box 1, RLGP, 4. An abridged version of this essay was published in 1902. See Garner, "In the Clutches of Cannibals."

41. McNutt, "Cannibals of Gabon."

42. On violence in the Nenguè-Nenguè region, see Bennett, "Ethnographic Notes on the Fang"; Chamberlin, "Competition and Conflict," 103–113.

43. Garner, "Among the Cannibals," 14–16.

44. Unless otherwise noted, the following is drawn from Garner, "Heathen Rites to Heathen Gods." Subsequent citations to this work are given in the text.

45. Du Chaillu, *Explorations and Adventures*, 493; Lewis, *Trader Horn*, 56.

46. Garner, "Heathen Rites," 17.

47. Garner, "Nkanjo," n.d., Nancy Bet folder, box 3, RLGP, 2.

48. Lenz, *Skizzen aus West-Afrika*, 61–62.

49. Gaulme, *Le pays de Cama*, 251–257.

50. Cinnamon, "Missionary Expertise"; Cinnamon, "Robert Hamill Nassau"; Rich, "Maurice Briault."

51. While there is ample literature regarding the dramatic rise of *bwiti* religious movements in the first half of the twentieth century, it does not take into account the burst of Catholic and Protestant religious conversions or the growth of other spiritual movements such as Madamoiselle. The gradual rise of the Christian and Missionary Alliance movement has yet to draw any scholarly research at all, despite its popularity in southern Gabon from the 1930s onward. On one missionary's perspective on Protestant revivals in the mid-1930s, see Perrier, *Gabon, un réveil religieux*. For a missionary review of the CMA's early history, see Thompson, *Au-déla des brumes*. The spread of Catholicism has not been well explored, but for a brief discussion, see Rich, "Marcel Lefebvre in Gabon."

52. Trial, *Okoume*; Dedet, *La mémoire du fleuve*.

53. Trial, *Okoume*, 245–256.

CHAPTER SEVEN. Aping Civilization

1. Richard Lowenberg, "Ape Story," accessed 5 February 2010, http://www.radlab.com/information/apestory.html.

2. Djuna Barnes, "The Girl and the Gorilla." *New York World*, 18 October 1914, 9.

3. Jones, "Gorilla Trails in Paradise"; Jones, "Missing Link."

4. Bradford and Blume, *Ota Benga*.

5. Haraway, *Primate Visions*, 26–58.

6. For example, see Kasson, *Perfect Man*, and Mayer, *Artificial Africas*, 48–75.

7. Amato, "White Elephant in London."

8. Morse, "Mark of the Beast"; Schnell, "Tiger Tales"; Leighton and Surridge, "Empire Bites Back"; Rothfels, "How the Caged Bird Sings."

9. Clark, *God or Gorilla*.

10. Ibid., 13.

11. Grier, *Pets in America*.

12. Hoganson, *Consumers' Imperium*.

13. "Monkey Intelligence," 2.

14. E. H. T, "Re-volution."

15. Andrews, "Woman's Council Table."

16. "Psychic Character of Apes," *Zion's Herald*, 23 August 1899, 34.

17. *Atlanta Journal-Constitution*, 2 February 1902, 16.

18. "Monkey Cotton Pickers," *Newark Daily Advocate [Newark, Ohio]*, 31 March 1899, 2.

19. "Monkey Cotton Pickers," *Fort Wayne News*, 25 April 1899, 13; "Monkey Cotton Pickers," *San Antonio Light*, 1 May 1899, 3.

20. Payn, "English Notes."

21. "A School for Monkeys," *Bruce Herald*, 26 April 1901, 2.

22. "Ape and Monkey Life: Reply by Taylor White," *Otago Witness*, 16 September 1903, 21.

23. Crowley, *Confessions of Aleister Crowley*, 746.

24. "Talks to Monkeys," *Van Wert Daily*, 1 September 1906, 6.

25. "What It Costs to Hunt Big Game in Africa a la Roosevelt," *Baltimore Sun*, 28 February 1909, 17.

26. "Lost to the World," *Los Angeles Times*, 4 August 1901, c3; "Adventurers Astray in Vast Wildernesses," *Pittsburgh Post-Gazette*, 3 August 1901, 15.

27. "A Defense of Mr. Hill," *Christian Union*, 11 June 1892, 45.

28. "Picked Up," *Atlanta Journal-Constitution*, 22 April 1893, 4.

29. *Atlanta Journal-Constitution*, 12 January 1893, 4; *Atlanta Journal-Constitution*, 5 September 1910, 4; *Atlanta Journal-Constitution*, 15 September 1910, 4.

30. "Twilight Thinks," *Pittsburgh Post-Gazette*, 31 August 1910, 6.

31. "The Farce of San Antonio," *Independent*, 13 April 1911, 805.

32. Uzawa, "White Man and Yellow Man."

33. The following material is drawn from "Togo Solves the Mystery of Republican Defeat," *New York Times*, 4 December 1910, SM 14.

34. Uzawa, "White Man and Yellow Man," 205–206.

35. Casson, "What Fools These Mortals Be," 716.

36. Casson, *Making Money Happily*, 5–6.

37. Bierce, "Civilization of the Monkey," 32–33, 35, 36.

38. "Monkey Business,"147.

39. "Sapient Simian," 28.

40. "Contemporary Humor," *Brooklyn Eagle*, 20 July 1893, 4.

41. "The Zoo Society Sees the Apes Dine," *New York Times*, 16 May 1911, 13.

42. Haraway, *Primate Visions*.

43. "Fact and Comment," *Youth's Companion*, 4 November 1915, 592.

44. "Big Gunda in Chains in Rage at Keeper," *New York Times*, 4 June 1915, 6.

45. Hancocks, "Zoo Animals as Entertainment Exhibitions," in Malamud, *Cultural History of Animals*, 103.

46. "Mme. Ningo," *Washington Post*, 30 September 1911, 6; "Zoo's Only Gorilla Dead," *New York Times*, 6 October 1911, 1.

47. "Civilization Too Much for Dinah Our Only Girl Gorilla," *Lima Daily News*, 18 September 1915, 3.

48. "Veterinary Surgeon in Zoological Park Must Be Versatile in His Methods," *Iowa City Press-Citizen*, 27 July 1923, 8.

49. "Jungle Baby Lolls in Invalid's Luxury," *New York Times*, 21 December 1914, 16.

50. "America's Only Gorilla Dies," *Washington Post*, 4 August 1915, 6.

51. Rothfels, "How the Caged Bird Sings," in Kete, *Cultural History of Animals*, 112.

52. Hornaday, *Minds and Manners*, 94.

53. "Civilization Too Much for Dinah," 3.

54. "Pines for Her Jungle," *Washington Post*, 27 September 1911, 6.

55. Mary Corthope, "Nostalgia, and Its Nepenthe," *San Antonio Light*, 17 November 1911, 6.

56. "Daddy's Bedtime Story: Dinah, the Gorilla, Takes an Airing," *La Crosse Tribune*, 19 January 1915, 2.

57. "This Lady Sits Tight on the Water Wagon," *Sioux County Herald*, 12 October 1911, 6.

58. "Hunger Strikers of the Bronx Zoo," *Fort Wayne Journal Gazette*, 22 October 1916, 23.

59. Kete, introduction to *Cultural History of Animals*, 15–17.

60. Eugenie Shonnard, interview by Sylvia Lewis, 27 February–9 April 1964,

Archives of American Art, Smithsonian Institution, accessed 10 February 2010, http://www.aaa.si.edu/collections/oralhistories/transcripts/shonna64.htm.

61. Loncraine, "Voix-de-Ville."

62. Barnes, "Girl and the Gorilla," 2.

63. Ibid.

64. Even though scholars intrigued by Barnes's early writings never discuss Garner, they have noted Barnes's destabilizing approach to male authority. See Larabee, "Early Attic Stage," in Broe, *Silence and Power*, 41; Diane Warren, *Djuna Barnes' Consuming Fictions*, 29.

CONCLUSION

1. Candland, *Feral Children and Clever Animals*, 207–226.

BIBLIOGRAPHY

Archives Consulted

France

Journaux de la Communauté de Sainte Anne de Fernan Vaz (JC). Archives of the Congregation of the Holy Spirit, Chevilly Larue, France.

Rapports Politiques, 1896–1920 (RP). Gouvernement Général de l'Afrique Équatoriale Française. Série 4(1)D1- D20. Centre d'Archives d'Outre-mer, Aix-en-Provence, France.

United States

Aschemeier, Charles. Field Notes, 1917–1918. Record Unit 105. Smithsonian Institution Archives, Washington, D.C.

Collins-Garner Expedition (CGE). Folders 9–12, Record Unit 45. Smithsonian Institution Archives Washington, D.C.

Garner, Richard Lynch. File (RLGF). Washington County Historical Society, Abingdon, Va.

Garner, Richard Lynch. Papers (RLGP). National Anthropological Archives. Smithsonian Institution, Washington, D.C.

Hornaday, William. Papers (WHP). Wildlife Conservation Society Library, New York.

Means, Ambrose. Papers. Private collection, Cathy Carter. Prescott, Ariz.

National Zoological Park (NZP). Folder 25. Record Unit 365. Smithsonian Institution Archives , Washington, D.C.

Selected Records of the War Department Related to Confederate Prisoners of War 1861–1865. Microfilm N598-39. U.S. War Department, National Archives and Records Administration, Washington, D.C.

Interviews

Dikoumba, Jean-Claude. 21 December 2008. Ndougou, Gabon.
Joussiant Nkombe, Pierre. 25 December 2008. Ombouè, Gabon.
Lamu, Pierre. 29 December 2008. Sainte Anne, Gabon.
Ndong, Nze. 30 December 2008. Assewe and Kongo, Gabon.
Ndong, Nze, Marie Nyi Nyondome, and Jeanne Eloghe. 30 December 2008. Kongo, Gabon.
Ngomba, Jean-Robert. 27 December 2008. Sainte Anne, Gabon.
Nkourou Ratowo, Henriette. 23 December 2008. Ombouè, Gabon.
Nkundi, Mikadi, and Pascal Nkundi. 19 July 2007. Biaka, Gabon.
Pesi, Celestine. 30 December 2008. Koungo, Gabon.
Yeno, Madeline. 27 December 2008. Sainte Anne, Gabon.

Journals and Newspapers Consulted

Africa Mail
Atlanta Journal-Constitution
Baltimore Sun
Brooklyn Eagle
Bruce Herald
Bulletin of the New York Zoological Society
Christian Union
Cosmopolitan
Fort Wayne Journal Gazette
Fort Wayne News
Hartford Courant
Independent
Iowa City Press-Citizen
La Crosse Tribune
Lima Daily News
Los Angeles Times
Mills County Tribune
Newark Daily Advocate
New York Times
New York Tribune
New York World
Oakland Tribune
Otago Witness

Pittsburgh Post-Gazette
Salt Lake Tribune
San Antonio Light
Sioux County Herald
Syracuse Evening Herald
Times of London
Town Topics
Van Wert Daily
Washington Post
Youth's Companion
Zion's Herald

Published Works

Adas, Michael. *Machines as the Measure of Men: Science, Technology, and Ideologies of Western Dominance.* Ithaca, N.Y.: Cornell University Press, 1989.

Agondjo-Okawe, Pierre-Louis. "Les droits fonciers coutumiers au Gabon (société Nkomi, groupe Myene)." *Revue Juridique et Politique: Indépendance et Cooperation* 24 (1970): 1135–1152.

———. "Structures parentales gabonaises et développement." Thesis, Faculté de Droit, University of Paris I, 1967.

Aleko, Hilaire, and Gilbert Peuch. "Notes sur la lagune Ngovè et les Ngubi." *Pholia* 3 (1988): 257–270.

Algeo, Katie. "Locals on Local Color: Imagining Identity in Appalachia." *Southern Cultures* 9, no. 4 (2003): 27–54.

Amato, Sarah. "The White Elephant in London: An Episode of Trickery, Racism, and Advertising." *Journal of Social History* 43, no. 1 (2009): 31–66.

Ambouroue Avaro, Joseph. *Un peuple Gabonais à l'aube de la colonization: Le Bas-Ogowe au XIXe siècle.* Paris: Karthala, 1981.

Ames, Eric. *Carl Hagenbeck's Empire of Entertainments.* Seattle: University of Washington Press, 2009.

Andrews, E. F. "Woman's Council Table: Cracker English." *Chautauquan*, April 1896, 1.

Angell, Stephen Ward. *Bishop Henry McNeal Turner and African-American Religion in the South.* Knoxville: University of Tennessee Press, 1992.

Appadurai, Arjun. "Commodities and the Politics of Value." In *The Social Life of Things: Commodities in a Cultural Perspective*, edited by Arjun Appadurai, 3–63. Cambridge: Cambridge University Press, 1986.

————. *The Social Life of Things: Commodities in a Cultural Perspective*. Cambridge: Cambridge University Press, 1986.

Aschemeier, Charles. "The Gorilla at Home." *Dearborn Independent*, 3 April 1926, 19–21.

————. "Life among the Natives of the Gorilla Country." *Dearborn Independent*, 21 May 1927, 21–23.

Baratay, Eric, and Elisabeth Hardouin-Fugier. *Zoo: A History of Zoological Gardens in the West*. London: Reaktion Books, 2002.

Barnes, Kenneth. *Journey of Hope: The Back-to-Africa Movement in Arkansas in the Late 1800s*. Chapel Hill: University of North Carolina Press, 2004.

Batteau, Allen. *The Invention of Appalachia*. Tucson: University of Arizona Press, 1990.

Bederman, Gail. *Manliness and Civilization: A Cultural History of Gender and Race in the United States, 1880–1917*. Chicago: University of Chicago Press, 1996.

Beinart, William, and JoAnn McGregor, eds. *Social History and African Environments*. Oxford: Currey, 2003.

Bennett, Albert. "Ethnographic Notes on the Fang." *Journal of the Royal Anthropological Institute of Great Britain and Ireland* 26 (1899): 66–98.

Bernault, Florence. "Body, Power, and Sacrifice in Equatorial Africa." *Journal of African History* 47 (2006): 207–239.

————. *Démocraties ambiguës en Afrique centrale: Gabon, Congo-Brazzaville, 1945–1965*. Paris: Karthala, 1996.

————. "Dévoreurs de la nation: Les migrations fang au Gabon." In *Être étranger et migrant en Afrique au XXe siècle*, edited by Catherine Coquery-Vidrovitch and Issiaka Mandé, 169–187. Paris: L'Harmattan, 2003.

Bierce, Ambrose. "Civilization of the Monkey." In *The Collected Works of Ambrose Bierce*, 32–36. Vol. 11. New York: Neale, 1911.

Billings, Dwight B., and Kathleen M. Blee. *The Road to Poverty: The Making of Wealth and Hardship in Appalachia*. New York: Cambridge University Press, 2000.

Billings, Dwight B., Mary Beth Pudup, and Altina Waller, eds. *Appalachia in the Making: The Mountain South in the Nineteenth Century*. Chapel Hill: University of North Carolina Press, 1995.

Blathwayt, Raymond. "Wild Animals: How They Are Captured, Transported, and Sold." *McClure's Magazine* June 1893, 27–33.

Blount, Alison. *Travel, Gender, and Imperialism: Mary Kingsley and West Africa*. New York: Guilford, 1994.

Bonhomme, Julien. *Le miroir et le crâne: Parcours initiatique du Bwete Mitsoko (Gabon)*. Paris: EHESS, 2006.

———. "Les tribulations de l'esprit blanc (et de ses merchandises): Voyages et aventures de Paul de Chaillu en Afrique équatoriale." *Cahiers d'Études Africaines* 183 (2006): 493–512.

Bourget, Marie-Noëlle, Christian Licoppe, and H. Ottom Sibum, eds. *Instruments, Travel and Science: Itineraries of Precision from the Seventeenth to the Twentieth Century.* New York: Routledge, 2002.

Bradford, Phillips Verner, and Harvey Blume. *Ota Benga: The Pygmy in the Zoo.* New York: St. Martin's Press, 1992.

Brandon-Jones, Christine. "Edward Blyth, Charles Darwin, and the Animal Trade in Nineteenth-Century India and Britain." *Journal of the History of Biology* 30, no. 2 (1997): 145–178.

Brantz, Dorothee. "The Domestication of Empire: Human-Animal Relationships at the Intersection of Civilization, Evolution, and Acclimatization in the Nineteenth Century." In Kete, *Cultural History of Animals*, 73–93.

Brechin, Gray. "Conserving the Race: Natural Aristocracies, Eugenics, and the U.S. Conservation Movement." *Antipode* 28 (1996): 229–245.

Brehm, Alfred. *Tierleben: Allegemeine Kunde des Tierreichs; Säugtiere.* Vol. 13. Leipzig, Germany: Bibliographisches Institut, 1916.

Bridges, William. *Gathering of Animals: An Unconventional History of the New York Zoological Society.* New York: Harper and Row, 1974.

Brisson, Ulrike. "Fish and Fetish: Mary Kingsley's Studies of Fetish in West Africa." *Journal of Narrative Theory* 35, no. 3 (2005): 326–340.

Brockington, Dan. *Fortress Conservation: The Preservation of the Mkomazi Game Reserve, Tanzania.* Bloomington: Indiana University Press, 2002.

Broe, Mary Lynn, ed. *Silence and Power: A Reevaluation of Djuna Barnes.* Carbondale: Southern Illinois University Press, 1991.

Bucher, Henry. "The Mpongwe of the Gabon Estuary: A History to 1860." PhD diss., University of Wisconsin, 1977.

Cadet, Xavier. *Histoire des Fang, peuple gabonais.* Paris: L'Harmattan, 2009.

Candland, Douglas Keith. *Feral Children and Clever Animals: Reflections on Human Nature.* New York: Oxford University Press, 1995.

"Can Monkeys Talk?" *American Review of Reviews* 42 (1910): 492.

Casson, Herbert. *Creative Thinkers: The Efficient Few Who Cause Progress and Prosperity.* New York: Forbes, 1929.

———. *Making Money Happily: Twelve Tips on Success and Happiness.* New York: Forbes, 1926.

———. "What Fools These Mortals Be." *Machinists' Monthly Journal,* November 1899, 716.

Caudill, Harry. *Night Comes to the Cumberlands: A Biography of a Depressed Area.* Boston: Little, Brown, 1963.

Challaye, Felicien. *Le Congo français*. Paris: Cahiers de la Quinzaine, 1906.

Chamberlin, Christopher. "Competition and Conflict: The Development of the Bulk Export Trade in Central Gabon during the 19th Century." PhD diss., University of California at Los Angeles, 1977.

Cheang, Sarah. "Women, Pets, and Imperialism: The British Pekingese Dog and Nostalgia for Old China." *Journal of British Studies* 45 (2006): 359–387.

Cinnamon, John. "The Long March of the Fang: Anthropology and History in Equatorial Africa." PhD diss., Yale University, 1998.

———. "Missionary Expertise, Social Science, and the Uses of Ethnographic Knowledge in Colonial Gabon." *History in Africa* 33 (2006): 413–432.

———. "Robert Hamill Nassau, Missionary Ethnography and the Colonial Encounter in Gabon." *Le Fait Missionnaire* 19 (2006): 37–64.

Clark, Constance. *God or Gorilla: Images of Evolution in the Jazz Age*. Baltimore: Johns Hopkins University Press, 2008.

"Communauté de Sainte Anne du Fernan Vaz." *Bulletin de la Congregation du Saint-Esprit* 24 (1905–1907): 209–215.

Conklin, Paul. *When All the Gods Trembled: Darwinism, Scopes, and American Intellectuals*. Lanham, Md.: Rowan and Littlefield, 1998.

Cookey, S. J. S. "The Concession Policy in the French Congo and the British Reaction, 1898–1906." *Journal of African History* 7 (1966): 263–278.

Cooper, Frederick. *Colonialism in Question: Theory, Knowledge, History*. Berkeley: University of California Press, 2005.

Coquery-Vidrovitch, Catherine. *Le Congo aux temps des grandes companies concessionaires, 1898–1930*. Paris: Plon, 1972.

Croke, Vicki. *The Modern Ark: The Story of Zoos Past, Present, and Future*. New York: Scribner, 1997.

Crowley, Aleister. *The Confessions of Aleister Crowley*. London: Penguin, 1969.

"Current Notes." *American Journal of Physical Anthropology* 3 (1920): 198.

Curtis, Sarah, and Kevin Callahan, eds. *Encountering French History*. Lincoln: University of Nebraska Press, 2008.

Dedet, Christian. *La mémoire du fleuve: L'Afrique aventureuse de Jean Michonnet*. Paris: Phébus, 1984.

Dehler, Gregory. "An American Crusader: William Temple Hornaday and Wildlife Protection in America, 1840–1940." PhD diss., Lehigh University, 2002.

Desmond, Adrian, and James Moore. *Darwin's Sacred Cause: How a Hatred of Slavery Shaped Darwin's Views on Human Evolution*. Boston: Houghton Mifflin Harcourt, 2009.

Ditmars, Richard. *Confessions of a Scientist*. New York: Macmillan, 1934.

Dotson, Rand. *Roanoke, Virginia, 1882–1912: Magic City of the New South*. Knoxville: University of Tennessee Press, 2007.

Dubois, Colette. "Le prix d'une guerre: Deux colonies pendant la première guerre mondiale (Gabon, Oubangui-Chari), 1911–1923." PhD diss., University of Aix-en-Provence, 1985.

Du Chaillu, Paul. *Explorations and Adventures in Equatorial Africa*. New York: Harper Brothers, 1861.

———. *A Journey to Ashango-Land*. New York: Appleton, 1867.

Duignan, Peter, and L. H. Gann. *The United States and Africa: A History*. New York: Cambridge University Press, 1984.

Dumoulin, Michel. *Léopold II, un roi génocidaire?* Brussels: Académie royale de Belgique, 2005.

Dunaway, Wilma. *Slavery in the Mountain South*. New York: Cambridge University Press, 2003.

Dunn, Arthur Wallace. *Gridiron Nights: Humorous and Satirical Views of Politics and Statesmen as Presented by the Famous Dining Club*. New York: Stokes, 1915.

Dupré, Georges. *Un ordre et sa destruction*. Paris: ORSTOM, 1982.

Early, Julie English. "Unescorted in Africa: Victorian Women Ethnographers Toiling in the Fields of Sensational Science." *Journal of American Culture* 18, no. 4 (1995): 67–95.

Easton, Robert, and Mackenzie Brown. *Lord of Beasts: The Saga of Buffalo Jones* (Tucson: University of Arizona Press, 1961.

E. H. T. "Re-volution." *Eclectic Magazine of Foreign Literature*, March 1893, 380.

Eller, Ronald D. *Miners, Millhands, and Mountaineers: Industrialization of the Appalachian South, 1880–1930*. Knoxville: University of Tennessee Press, 1982.

Fabian, Johannes. *Out of Our Minds: Reason and Madness in the Exploration of Central Africa*. Berkeley: University of California Press, 2000.

Fairhead, James, and Melissa Leach. *Misreading the African Landscape: Society and Ecology in a Forest-Savanna Mosaic*. Cambridge: Cambridge University Press, 1996.

Fay, J. Michael, and Michael Nicholas. *The Last Place on Earth*. Washington, D.C.: National Geographic Society, 2005.

Fernandez, James. *Bwiti: An Ethnography of the Religious Imagination in Africa*. Princeton: Princeton University Press, 1982.

Fishel Leslie H., Jr., and Benjamin Quarles. *The Black American*. Glenview, Ill.: Scott and Foresman, 1970.

Fitzgerald, Michael. "'We Have Found a Moses': Theodore Bilbo, Black Nationalism, and the Greater Liberia Bill of 1939." *Journal of Southern History* 63, no. 2 (1997): 293–320.

Fleeger, Robert. "Theodore G. Bilbo and the Decline of Public Racism, 1938–1947." *Journal of Mississippi History* 68 (2007): 1–27.

Fôret, Auguste. "Le lac Fernan Vaz." *Bulletin de la Société de Géographie de Paris* (1898): 308–327.

Fredrickson, George M. *The Black Image in the White Mind: The Debate on Afro-American Character and Destiny, 1817–1914.* New York: Harper and Row, 1971.

Friend, Craig Thompson, ed. "From Southern Manhood to Southern Masculinities." In Friend, *Southern Masculinity*, xv.

———. *Southern Masculinity: Perspectives on Manhood in the South since Reconstruction.* Athens: University of Georgia Press, 2009.

Friend, Craig Thompson, and Lorri Glover, eds. *Southern Manhood: Perspectives on Masculinity in the Old South.* Athens: University of Georgia Press, 2004.

Gardinier, David. "The American Board (1842–1870) and Presbyterian Board (1870–1892) Missions in Northern Gabon and African Responses." *Africana Journal* 17 (1998): 215–234.

———. "The American Presbyterian Mission in Gabon: Male Mpongwe Converts and Agents." *Journal of Presbyterian History* 69 (1991): 61–70.

Garland, Elizabeth. "The Elephant in the Room: Confronting the Colonial Character of Wildlife Conservation in Africa." *African Studies Review* 51, no. 3 (2008): 51–74.

Garner, Richard L. "Adventures in Central Africa: Domestic Arrangements in the Jungle." *Century* 99 (March 1920): 595–604.

———. "Adventures in Central Africa: Hunt and Wild Life." *Century* 99 (May 1920): 125–135.

———. "Adventures in Central Africa: Peculiar Monkeys and Antelopes." *Century* 99 (April 1920): 842–852.

———. "Among the Gorillas: A Voice from the Wilderness." *McClure's Magazine* 1 (1893): 364–372.

———. *Apes and Monkeys.* Boston: Ginn, 1900.

———. *Autobiography of a Boy.* Washington, D.C.: Huff Duplicating, 1930.

———. "Contemporary Ancestors of Ours." *Independent*, 10 January 1920, 60–61.

———. "A Hermit's Home." *Modern Women* (October 1906): 110, 120.

———. "Hunting Big Game in Africa." *Sports Afield* (September 1907): 396–405.

———. "In the Clutches of Cannibals." *World Wide Magazine* 9 (1902): 597–603.

———. "My Recent Work, and Susie." *Independent*, 8 September 1910, 518–523.

———. "Native Institutions of the Ogowe Tribes of West Central Africa: An Interpretation of Their Meaning as Viewed from the Standpoint of the Native Philosopher." *Journal of the African Society* 1, no. 3 (1902): 369–380.

———. "Nyanga, the Queer Boy." *American Boy* (February 1920): 15–16, 46.

————. "The Religion of African Cannibals." *Forum* 63 (1920): 305–314.

————. *The Speech of Monkeys*. New York: Webster, 1892.

————. "What I Expect to Do in Africa." *North American Review* 1541 (1892): 713–718.

————. "Why I Know That Monkeys Talk." *Popular Science Monthly* 96, no. 4 (1920): 21–23.

Gaulme, François. "Le Bwiti chez les Nkomi: Association cultuelles et évolution historique sur le littoral gabonais." *Journal des Africanistes* 49, no. 2 (1979): 37–87.

————. "Paul Du Chaillu et les Nkomi ou l'explorateur participant." In Hombert and Perrois, *Coeur d'Afrique*, 146–161.

————. *Le pays de Cama: Un ancient État côtier du Gabon et ses origines*. Paris: Karthala, 1981.

————. "Un problème d'histoire du Gabon: Le sacre du Père Bichet par les Nkomi en 1897." *Revue Française d'Histoire d'Outre-Mer* 61 (1974): 395–416.

Geschiere, Peter. *The Modernity of Witchcraft*. Charlottesville: University of Virginia Press, 1997.

————. "Tournaments of Value in the Forest Area of Southern Cameroon: 'Multiple Self-Realization' versus Colonial Coercion during the Rubber Boom (1900–1913)." In *Commodification: Things, Agency, and Identities*, edited by Peter Geschiere and Wim M. J. van Binsbergen , 243–265. Munster, Germany: LIT, 2005.

Geschiere, Peter, and Wim M. J. van Binsbergen, eds. *Commodification: Things, Agency, and Identities*. Munster, Germany: LIT, 2005.

Gilmore, Glenda Elizabeth. *Gender and Jim Crow: Women and the Politics of White Supremacy in North Carolina, 1896–1920*. Chapel Hill: University of North Carolina, 1996.

Gordon, Larry. *The Last Confederate General: John C. Vaughn and His East Tennessee Calvary*. Minneapolis: Zenith, 2008.

Gordon, Richard. "Fido: Dog Tales of Colonialism in Namibia." In *Social History and African Environments*, edited by William Beinart and JoAnn McGregor, 240–254. Oxford: Currey, 2003.

Gray, Christopher. *Colonial Rule and Crisis in Equatorial Africa: Southern Gabon, ca. 1850–1940*. Rochester, N.Y.: University of Rochester Press, 2002.

Grier, Katherine. *Pets in America: A History*. Chapel Hill: University of North Carolina Press, 2006.

Guterl, Matthew Pratt. *The Color of Race in America, 1900–1940*. Cambridge, Mass.: Harvard University Press, 2001.

Guyer, Jane, and Samuel-Martin Eno Belinga. "Wealth in People as Wealth in

Knowledge: Accumulation and Composition in Equatorial Africa." *Journal of African History* 36 (1995): 91–120.

Hancocks, David. *A Different Nature: The Paradoxical World of Zoos and Their Uncertain Future*. Berkeley: University of California Press, 2001.

———. "Zoo Animals as Entertainment Exhibitions." In Malamud, *Cultural History of Animals*, 95–118.

Hanson, Elizabeth Anne. *Animal Attractions: Nature on Display in American Zoos*. Princeton, N.J.: Princeton University Press, 2002.

Haraway, Donna. *Primate Visions: Gender, Race, and Nature in the World of Modern Science*. New York: Routledge, 1989.

Harper, Lila Marz. *Solitary Travelers: Nineteenth-Century Women's Travel Narratives and the Scientific Vocation*. Madison, N.J.: Fairleigh Dickinson University Press, 2001.

Herzig, Rebecca. *Suffering for Science: Reason and Sacrifice in Modern America*. Ithaca, N.Y.: Cornell University Press, 2005.

Hickey, Dennis, and Kenneth C. Wylie. *An Enchanting Darkness: The American Vision of Africa in the Twentieth Century*. East Lansing: Michigan State University Press, 1993.

Hochshild, Adam. *King Leopold's Ghost: A Story of Greed, Terror, and Heroism in Colonial Africa*. New York: Mariner, 1998.

Hoganson, Kristen. *Consumers' Imperium: The Global Production of American Domesticity, 1865–1920*. Chapel Hill: University of North Carolina Press, 2007.

———. *Fighting for American Manhood: How Gender Politics Provoked the Spanish-American and Philippine-American Wars*. New Haven: Yale University Press, 1998.

Hombert, Jean-Marie, and Louis Perrois, eds. *Coeur d'Afrique: Gorilles, cannibales, et pygmées dans le Gabon de Paul Du Chaillu*. Paris: CNRS, 2007.

Hornaday, William. "A Bird's-Eye View of the Animal Kingdom." *St. Nicholas: A Monthly Magazine for Boys and Girls* 21 (1893–1894): 231–237.

———. "Department of Mammals." *Annual Report of the New York Zoological Society* 37 (1914): 70–73.

———. *The Minds and Manners of Wild Animals*. New York: Scribner's Sons, 1922.

Hsiung, David. *Two Worlds in the Tennessee Mountains: Exploring the Origins of Appalachia*. Lexington: University Press of Kentucky, 1997.

"An Intelligent Chimpanzee." *St. Nicholas*, June 1912, 750.

Israel, Charles. *Before Scopes: Evangelicalism, Education, and Evolution in Tennessee, 1870–1925*. Athens: University of Georgia Press, 2004.

Jackson, John, Jr. *Science, Race, and the Case against Brown v. Board of Education*. New York: New York University Press, 2005.

Jacobs, Nancy. "The Intimate Politics of Ornithology in Colonial Africa." *Comparative Studies in Society and History* 48 (2006): 564–603.

Jacobson, Matthew Frye. *Barbarian Virtues: The United States Encounters Foreign Peoples at Home and Abroad, 1876–1917.* New York: Macmillan, 2001.

Jacoby, Susan. *Freethinkers: A History of American Secularism.* New York: Metropolitan, 2004.

Jean-Baptiste, Rachel. "Une ville libre? Marriage, Divorce, and Sexuality in Colonial Libreville, Gabon, 1849–1960." PhD diss., Stanford University, 2005.

Jones, Jeannette Eileen. "'Gorilla Trails in Paradise': Carl Akeley, Mary Bradley, and the American Search for the Missing Link." *Journal of American Culture* 29, no. 3 (2006): 321–336.

———. *In Search of Brightest Africa: Imagining Africa in America, 1884–1936.* Athens: University of Georgia Press, 2010.

———. "Simians, Negroes, and the 'Missing Link': Evolutionary Discourses and Transatlantic Debates on 'The Negro Question.'" In *Darwin in Atlantic Cultures: Evolutionary Visions of Race, Gender, and Sexuality,* edited by Jeannette Eileen Jones and Patrick Sharp, 191–207. New York: Routledge, 2009.

Jones, Jeannette Eileen, and Patrick Sharp, eds. *Darwin in Atlantic Cultures: Evolutionary Visions of Race, Gender, and Sexuality.* New York: Routledge, 2009.

Kaplan, Amy. *The Anarchy of Empire in the Making of U.S. Culture.* Cambridge, Mass.: Harvard University Press, 2002.

Kasson, John. *Houdini, Tarzan, and the Perfect Man: The White Male Body and the Challenge of Modernity in America.* New York: Hill and Wang, 2001.

Keim, Curtis. *Mistaking Africa: Curiosities and Inventions of the American Mind.* 2nd ed. Boulder, Colo.: Westview, 2007.

Kennedy, Dane. *Islands of White: Settler Society in Kenya and Southern Rhodesia.* Durham, N.C.: Duke University Press, 1987.

Kete, Kathleen, ed. *A Cultural History of Animals in the Age of Empire.* New York: Berg, 2007.

Laburthe-Tolra, Pierre. *Les seigneurs de la fôret: Essai sur le passé historique, l'organisation sociale et les norms éthiques des anciens Beti du Cameroun.* Paris: Sorbonne, 1981.

Lake, Marilyn, and Henry Reynolds. *Drawing the Global Colour Line: White Men's Countries and the International Challenge of Racial Equality.* New York: Cambridge University Press, 2008.

Larabee, Ann. "The Early Attic Stage of Djuna Barnes." In Broe, *Silence and Power,* 37–45.

Larson, Edward. *Summer for the Gods: The Scopes Trial and America's Continuing Debate over Science and Religion.* New York: Basic Books, 1997.

Larson, Orvin. *American Infidel: Robert G. Ingersoll.* New York: Citadel, 1962.

Leighton, Mary Elizabeth, and Lisa Surridge. "The Empire Bites Back: The Racialized Crocodile of the Nineteenth Century." In *Victorian Animal Dreams: Representations of Animals in Victorian Literature and Culture*, edited by Deborah Denenholz Morse and Martin Danahay, 249–270. Burlington, Vt.: Ashgate, 2007.

Lenz, Oskar. *Skizzen aus West-Afrika.* Berlin: Hofmann, 1878.

Leopold, Robert. *A Guide to Early Africa Collections in the Smithsonian Collection.* Washington, D.C.: Smithsonian Institution, 1994.

Lewis, Ethelreda, ed. *Trader Horn: Being the Life and Works of Alfred Aloysius Horn.* New York: Literary Guild of America, 1927.

Lewis, Helen M., Linda Johnson, and Donald Askins, eds. *Colonialism in Modern America: The Appalachian Case.* Boone, N.C.: Appalachian Consortium, 1978.

Lienesch, Michael. *In the Beginning: Fundamentalism, the Scopes Trial, and the Making of the Antievolution Movement.* Chapel Hill: University of North Carolina, 2007.

Loncraine, Rebecca. "Voix-de-Ville: Djuna Barnes' Stunt Journalism, Harry Houdini, and the Birth of Cinema." *Women: A Cultural Review* 19 (2008): 156–171.

Loungou Mouele, Théophile. "Le Gabon de 1910 à 1925: Les incidences de la première guerre mondiale sur l'évolution politique, économique et sociale." PhD diss., University of Aix-en-Provence, 1984.

Love, Eric J. T. *Race over Empire: Racism and U.S. Imperialism, 1865–1900.* Chapel Hill: University of North Carolina Press, 2004.

Lyon, Peter. *Success Story: The Life and Times of S. S. McClure.* New York: Scribner's Sons, 1963.

MacGaffey, Wyatt. *Kongo Political Culture.* Bloomington: Indiana University Press, 2000.

Mackenzie, John. *The Empire of Nature: Hunting, Conservation, and the English Empire.* Manchester, England: Manchester University Press, 1988.

———, ed. *Imperialism in the Natural World.* Manchester, England: Manchester University Press, 1990.

Malamud, Randy, ed. *A Cultural History of Animals in the Modern Age.* New York: Berg, 2007.

Mandelstam, Joel. "Du Chaillu's Stuffed Gorillas and the Savants from the British Museum." *Notes and Records of the Royal Society of London* 48, no. 2 (1994): 227–245.

Martin, Jean. *Savorgnan de Brazza.* Paris: Indes Savantes, 2005.

Martin, Phyllis. *Leisure and Society in Colonial Brazzaville*. Cambridge: Cambridge University Press, 1995.

Mary, André. *Le défi du syncrétisme: Le travail symbolique de la religion d'Eboga, Gabon*. Paris: EHESS, 1999.

Mayer, Ruth. *Artificial Africas: Colonial Images in the Times of Globalization*. Hanover, N.H.: University Press of New England, 2002.

Mays, Thomas. *The Saltville Massacre*. Abeline, Tex.: McWhiney Foundation Press, 1998.

M'Bokolo, Elikia. *Noirs et blancs en Afrique équatoriale: Les societes côtières et la pénétration française (vers 1820–1874)*. Paris: Mouton, 1981.

McCook, Stuart. "'It May Be Truth, but It Is Not Evidence': Paul du Chaillu and the Legitimation of Evidence in the Field Sciences." *Osiris* 11 (1996): 177–197.

McEwan, Cheryl. *Gender, Geography, and Empire: Victorian Women Travelers in Africa*. Burlington, Vt.: Ashgate, 2000.

McKnight, David. *Contested Borderland: The Civil War in Appalachian Kentucky and Virginia*. Lexington: University Press of Kentucky, 2006.

McNutt, P. A. "Cannibals of Gabon." *World Wide Magazine* 3 (1899): 403–412.

Mills, Sara. *Discourses of Differences: An Analysis of Women's Travel Writing and Colonialism*. New York: Routledge, 1991.

"Monkey Business." *Life*, 21 February 1901, 147.

Moran, Jeffrey. "The Scopes Trial and Southern Fundamentalism in Black and White: Race, Region, and Religion." *Journal of Southern History* 70, no. 1 (2004): 95–120.

Morel, Edmund. D. *The British Case in French Congo*. London: Heinemann, 1903.

Morse, Deborah Denenholz. "'The Mark of the Beast': Animals as Sites of Imperial Encounter from *Wuthering Heights* to *Green Mansions*." In *Victorian Animal Dreams: Representations of Animals in Victorian Literature and Culture*, edited by Deborah Denenholz Morse and Martin Danahay, 181–200. Burlington, Vt.: Ashgate, 2007.

Morse, Deborah Denenholz, and Martin Danahay, eds. *Victorian Animal Dreams: Representations of Animals in Victorian Literature and Culture*, Burlington, Vt.: Ashgate, 2007.

Murray, Narisara. "Lives of the Zoo: Charismatic Animals in the Social Worlds of the Zoological Gardens of London 1850–1897." PhD diss., Indiana University, 2004.

"Museum Donations." *Calendar of the University of Toronto and University College, 1896–1897*. Toronto: University of Toronto, 1897.

"Museum Notes." *American Museum Journal* 15 (1915): 375–376.

Musselman, Elizabeth Green. "Plant Knowledge at the Cape: A Study in African and European Collaboration." *International Journal of African Historical Studies* 36, no. 1 (2003): 367–392

Nettels, Elsa. *Language, Race, and Social Class in Howells's America.* Lexington: University Press of Kentucky, 1988.

Neumann, Roderick. *Imposing Wilderness: Struggles over Livelihood and Nature Preservation in Africa.* Berkeley: University of California Press, 1998.

Newby, Idus. *Jim Crow's Defense: Anti-Negro Thought in America, 1900–1930.* Baton Rouge: Louisiana State University Press, 1965.

Newell, Stephanie. *The Forger's Tale: In Search of Odeziaku.* Athens: Ohio University Press, 2006.

Noe, Kenneth. *Southwest Virginia's Railroad: Modernization and the Sectional Crisis in the Civil War Era.* Tuscaloosa: University of Alabama Press, 2003.

Numbers, Ronald. *Darwinism Comes to America.* Cambridge, Mass.: Harvard University Press, 1998.

Numbers, Ronald, and John Stenhouse, eds. *Disseminating Darwinism: The Role of Place, Race, Religion, and Gender.* New York: Cambridge University Press, 2001.

Numbers, Ronald, and Lester Stephens. "Darwinism in the American South." In Numbers and Stenhouse, *Disseminating Darwinism,*123–144.

Ó Síocháin, Séamas, and Michael O'Sullivan, eds. *The Eyes of Another Race: Roger Casement's Congo Report and 1903 Diary.* Dublin: University College of Dublin Press, 2004.

Parrish, Susan Scott. *American Curiosity: Cultures of Natural History in the Colonial British Atlantic World.* Chapel Hill: University of North Carolina Press, 2006.

Patterson, K. David. *The Northern Gabonese Coast to 1875.* Oxford: Oxford University Press, 1975.

———. "Paul B. Du Chaillu and the Exploration of Gabon, 1855–1865." *International Journal of African Historical Studies* 7, no. 4 (1974): 647–667.

———. "The Vanishing Mpongwe: European Contact and Demographic Change in the Gabon River." *Journal of African History* 16 (1975): 217–238.

Payn, James. "English Notes." *Independent,* 30 January 1896, 2.

Perrier, André. *Gabon, un réveil religieux.* Paris: L'Harmattan, 1988.

Pettegrew, Bruce. *Brutes in Suits: Male Sensibility in America, 1890–1920.* Baltimore: Johns Hopkins University Press, 2007.

Pratt, Mary Louise. *Imperial Eyes: Travel Writing and Transculturation.* New York: Routledge, 1992.

Prestholdt, Jeremy. *Domesticating the World: African Consumerism and the Genealogies of Globalization.* Berkeley: University of California Press, 2008.

Preston, J. "Educating a Chimpanzee." *Technical World* 15 (1911): 424–425.

Rabenkogo, Nicaise. "Le littoral du Nkomi (Gabon): Contribution géographique à la conservation des milieu naturels." PhD diss., Université Toulouse III, 2007.

Radick, Gregory. *The Simian Tongue: The Long Debate about Animal Language.* Chicago: University of Chicago Press, 2007.

Rand, Austin, Herbert Friedmann, and Melvin Traylor. "Birds from Gabon and Moyen Congo." *Fieldiana: Zoology* 41, no. 2 (1959): 219–411.

Raponda Walker, André. *Contes Gabonaises.* Paris: Présence Africaine, 1967.

———. *Notes d'histoire du Gabon.* 2nd ed. 1960. Reprint, Libreville, Gabon: Fondation Raponda Walker, 1996.

Reade, Winwood. *Savage Africa: Being the Narrative of a Tour in Equatorial Southwestern and Northwestern Africa.* London: Harper and Brothers, 1864.

Redkey, Edwin S. *Black Exodus: Black Nationalist and Back-to-Africa Movements, 1890–1910.* New Haven, Conn.: Yale University Press, 1969.

Renoff, Gregory. *The Big Tent: The Traveling Circus in Georgia, 1820–1930.* Athens: University of Georgia Press, 2008.

Rich, Jeremy. "After the Slave Trade, the Sea Remains: Mobility and Atlantic Networks in Gabon, ca. 1860–1920." *Atlantic Studies* 6 (2007): 153–172.

———. "Une Babylone Noire: Interracial Unions in Colonial Libreville, c. 1870–1914." *French Colonial History* 4 (2003): 145–170.

———. "Chimpanzees in the Colonial Maelstrom: Struggles over Knowledge, Race, and Commodities in the Gabonese Primate Trade, c. 1850–1940." In *Landscapes and Environments in Colonial and Post-Colonial Africa,* edited by Toyin Falola and Emily Brownell. New York: Routledge, forthcoming.

———. "Civilized Attire: Dress, Cultural Change and Status in Libreville, Gabon, ca. 1860–1914." *Cultural and Social History* 2, no. 2 (2005): 189–214.

———. "Cruel Guards and Anxious Chiefs: Fang Masculinities and State Power in the Gabon Estuary, 1920–1960." *Cahiers d'Etudes Africaines* 195 (2009): 705–732.

———. "Gabonese Men for French Decency: The Rise and Fall of the Gabonese *Ligue des Droits de l'Homme,* 1916–1939." *French Colonial History* 12 (forthcoming).

———. "King or Knave? Felix Adende Rapontchombo and Political Survival in the Gabon Estuary." *African Studies Quarterly* 6, no. 3 (2002), http://www.web.africa.ufl.edu/asq/v6/v6i3a1.htm.

———. "Marcel Lefebvre in Gabon: Revival, Missionaries, and the Colonial Roots of Catholic Traditionalism." In *Encountering French History,* edited by Sarah Curtis and Kevin Callahan, 53–83. Lincoln: University of Nebraska Press, 2008.

———. "Maurice Briault, André Raponda Walker and the Value of Missionary Anthropology in Colonial Gabon." *Le Fait Missionnaire* 19 (2006): 65–89.

———. "My Matrimonial Bureau: Masculine Concerns and Presbyterian Mission Evangelization in the Gabon Estuary, ca. 1900–1915." *Journal of Religion in Africa* 36, no. 2 (2006): 200–223.

———. "White Coronations and Magical Boycotts: Omyènè Political Strategies, Clan Leaders, and French Rule in Coastal Gabon, 1870–1920." *International Journal of African Historical Studies* 43, no. 2 (2010): 207–226.

———. *A Workman Is Worthy of His Meat: Food and Colonialism in the Gabon Estuary.* Lincoln: University of Nebraska Press, 2007.

Ritvo, Harriet. *The Animal Estate: The English and Other Creatures in the Victorian Age.* Cambridge, Mass.: Harvard University Press, 1987.

Rothfels, Nigel. "How the Caged Bird Sings: Animals and Entertainment." In Kete, *Cultural History of Animals*, 95–112.

———. *Savages and Beasts: The Birth of the Modern Zoo.* Baltimore: Johns Hopkins University Press, 1998.

Rotundo, E. Anthony. *American Manhood: Transformations in Masculinity from the Revolution to the Modern Era.* New York: Basic Books, 1994.

Safier, Neil. "Fruitless Botany: Joseph de Jussieu's South American Odyssey." In *Science and Empire in the Atlantic World*, edited by James Delbourgo and Nicholas Dew, 203–224. New York: Routledge, 2007.

———. *Measuring the New World: Enlightenment Science and South America.* Chicago: University of Chicago Press, 2008.

"The Sapient Simian." *Life*, 5 January 1911, 28.

Savage, Thomas, and Jeffries Wyman. "Notes of the External Characters and Habits of Troglodytes Gorilla: A New Species of Orang from the Gaboon River." *Boston Journal of Natural History* 5 (1847): 245–247.

Schaffer, Simon. "Golden Means: Assay Instruments and the Geography of Precision in the Guinea Trade." In *Instruments, Travel and Science: Itineraries of Precision from the Seventeenth to the Twentieth Century*, edited by Marie-Noëlle Bourguet, Christian Licoppe, and H. Ottom Sibum, 20–50. New York: Routledge, 2002.

Schiebinger, Londa. *Plants and Empire: Colonial Bioprospecting in the Atlantic World.* Cambridge, Mass.: Harvard University Press, 2004.

Schiebinger, Londa, and Claudia Swan, eds. *Colonial Botany: Science, Commerce, and Politics.* Philadelphia: University of Pennsylvania Press, 2005.

Schnell, Heather. "Tiger Tales." In *Victorian Animal Dreams: Representations of Animals in Victorian Literature and Culture*, edited by Deborah Denenholz Morse and Martin Danahay, 229–248. Burlington, Vt.: Ashgate, 2007.

Shapiro, Henry. *Appalachia on Our Mind: Southern Mountains and Mountaineers in the American Consciousness, 1870–1920*. Chapel Hill: University of North Carolina Press, 1978.

Silber, Nina. *The Romance of Reunion: Northerners and the South, 1865–1900*. Chapel Hill: University of North Carolina Press, 1993.

Sillans, Roger, and André Raponda Walker. *Rites et croyances des peoples du Gabon*. Paris: Presence Africaine, 1962.

Sinha, Mrinalini. *Colonial Masculinity: The "Manly Englishman" and the "Effeminate Bengali" in the Late Nineteenth Century*. Manchester, England: Manchester University Press, 1995.

Skabelund, Aaron. "Breeding Racism: The Imperial Battlefields of the 'German' Shepherd." *Society and Animals* 16 (2008): 354–371.

———. "Fascism's Furry Friends: Dogs, National Identity, and Racial Purity in 1930s Japan." In *The Culture of Japanese Fascism*, edited by Alan Tansman, 155–184. Durham, N.C.: Duke University Press, 2009.

Slade, Ruth. *King Leopold's Congo*. London: Oxford University Press, 1963.

Smith, Charles Spencer. *Glimpses of Africa, West and Southwest Coast*. Nashville, Tenn.: A. M. E. Sunday School Union, 1895.

Sondage, Scott. *Born Losers: A History of Failure in America*. Cambridge, Mass.: Harvard University Press, 2005.

Spiro, Jonathan P. *Defending the Master Race: Conservation, Eugenics, and the Legacy of Madison Grant*. Burlington: University of Vermont Press, 2009.

Stecopoulos, Harilaos. *Reconstructing the World: Southern Fictions and U.S. Imperialisms, 1898–1975*. Ithaca, N.Y.: Cornell University Press, 2008.

Stein, Judith. *The World of Marcus Garvey: Race and Class in Modern Society*. Baton Rouge: Louisiana University Press, 1991.

Steinhart, Ian. *Black Poachers, White Hunters: A Social History of Hunting in Colonial Kenya*. Athens: Ohio University Press, 2006.

Stern, Alexandra Minna. *Eugenic Nation: Faults and Frontiers of Better Breeding in Modern America*. Berkeley: University of California Press, 2005.

Storey, William. "Big Cats and Imperialism: Lion and Tiger Hunting in Kenya and Northern India, 1898–1930." *Journal of World History* 2 (1991): 135–173.

Summers, Lewis Preston. *History of Southwestern Virginia, 1746–1786 and Washington County, 1777–1870*. Richmond, Va.: Hill, 1903.

Sundiata, Ibrahim. *Brothers and Strangers: Black Zion, Black Slavery, 1914–1940*. Durham, N.C.: Duke University Press, 2003.

Swiderski, Stanislaw. *La religion Bouiti*. 5 vols. Ottawa, Ontario: Legas, 1990.

Tansman, Alan, ed. *The Culture of Japanese Fascism*. Durham, N.C.: Duke University Press, 2009.

Thomas, Keith. *Man and the Natural World: Changing Attitudes in England, 1500–1800*. New York: Oxford University Press, 1983.

Thompson, David. *Au-déla des brumes: L'histoire de l'amour de Dieu pour le peuple gabonais*. Libreville, Gabon: Librarie de l'Alliance Chrétienne, 2004.

Tillinghast, James Alexander. *The Negro in Africa and America*. New York: American Economic Association, 1902.

Tonda, Joseph. *Le Souverain modern: Le corps du pouvoir en Afrique central*. Paris: Karthala, 2005.

"Town Topics." *Town Topics* 35, no. 8 (20 September 1896): 13.

Trial, Georges. *Okoume*. Paris: Michel, 1939.

Tropp, Jacob. "Dogs, Poison, and the Meaning of Colonial Intervention in the Transkei, South Africa." *Journal of African History* 43 (2002): 451–472.

Turner, James. *Without God, without Creed: The Origins of Unbelief in America*. Baltimore: Johns Hopkins University Press, 1985.

Uzawa, Yoshiko. "'Will White Man and Yellow Man Ever Mix?': Wallace Irwin, Hashimura Togo, and the Japanese Immigrant in America." *Japanese Journal of American Studies* 17 (2006): 201–222.

Vande weghe, Jean-Pierre. *Les parcs nationaux du Gabon: Loango, Mayumba et le bas Ogooué*. Libreville, Gabon: Wildlife Conservation Society, 2007.

Vansina, Jan. *Paths in the Rainforest*. Madison: University of Wisconsin Press, 1990.

Van Sittert, Lance, and Sandra Swart, eds. *Canis Familiaris: A Dog History of South Africa*. Leiden, Holland: Brill, 2008.

Vanthemsche, Guy. *La Belgique et le Congo*. Brussels, Belgium: Complexe, 2007.

Veistroffer, Albert. *Vingt ans dans la brousse africaine*. Lille, France: Mercurs de Flandre, 1931.

Ward, Jason Morgan. "'A Richmond Institution': Earnest Sevier Cox, Racial Propaganda, and Organized Opposition to the Civil Rights Movement." *Virginia Magazine of History and Biography* 116 (September 2008): 262–293.

Warren, Diane. *Djuna Barnes' Consuming Fictions*. Burlington, Vt.: Ashgate, 2008.

Warren, Sidney. *American Freethought, 1860–1914*. 2nd ed. 1943. Reprint, New York: Gordian, 1966.

Welch, Richard E., Jr. *Response to Imperialism: The United States and the Philippine-American War*. Chapel Hill: University of North Carolina Press, 1979.

West, Stephen A. *From Yeoman to Redneck in the South Carolina Upcountry, 1850–1915*. Columbia: University of South Carolina Press, 2008.

Whitehead, Fred, and Verle Muhrer, eds. *Freethought on the American Frontier*. Buffalo, N.Y.: Prometheus Books, 1992.

Wylie, Diana. *Starving on a Full Stomach: Hunger and the Triumph of Cultural*

Racism in Modern South Africa. Charlottesville: University of Virginia Press, 2001.

Yerkes, Robert, and Ada Yerkes. *The Great Apes: A Study of Anthropoid Life.* New Haven, Conn.: Yale University Press, 1929.

Zimmerman, Andrew. "'What do you really want in German East Africa, Herr Professor?' Counterinsurgency and the Science Effect in Colonial Tanzania." *Comparative Studies in Society and History* 48 (2006): 419–461.

RACE IN THE ATLANTIC WORLD, 1700–1900

CPSIA information can be obtained at www.ICGtesting.com
Printed in the USA
LVOW060346231111

256145LV00004B/2/P